THE
SLUMBERING
MASSES

The publication of this book was assisted by
a bequest from Josiah H. Chase to honor his parents,
Ellen Rankin Chase and Josiah Hook Chase,
Minnesota territorial pioneers.

THE SLUMBERING MASSES

Sleep, Medicine, and Modern American Life

Matthew J. Wolf-Meyer

A QUADRANT BOOK

University of Minnesota Press
Minneapolis
London

Quadrant, a joint initiative of the University of Minnesota Press and the Institute for Advanced Study at the University of Minnesota, provides support for interdisciplinary scholarship within a new, more collaborative model of research and publication.

QUADRANT

Sponsored by Quadrant's Health and Society group (advisory board: Susan Craddock, Jennifer Gunn, Alex Rothman, and Karen-Sue Taussig).

http://www.quadrant.umn.edu

An earlier version of chapter 2 was published as "The Nature of Sleep," *Comparative Studies in Society and History* 53, no. 4 (2011): 945–70; copyright 2011 Society for the Comparative Study of Society and History; reprinted with permission of Cambridge University Press. An earlier version of chapter 3 was published as "Sleep, Signification, and the Abstract Body of Allopathic Medicine," *Body and Society* 14, no. 4 (December 2008): 93–114. Earlier versions of chapters 4 and 6 were published as "Precipitating Pharmakologies and Capital Entrapments: Narcolepsy and the Strange Cases of Provigil and Xyrem," *Medical Anthropology* 28, no. 1 (February 2009): 11–30; reprinted by permission of Taylor and Francis, Ltd. An earlier version of chapter 12 was published as "Fantasies of Extremes: Sports, War, and the Science of Sleep," *Biosocieties* 4, no. 2 (August 2009): 257–71.

Published by the University of Minnesota Press
111 Third Avenue South, Suite 290
Minneapolis, MN 55401-2520
http://www.upress.umn.edu

Library of Congress Cataloging-in-Publication Data

Wolf-Meyer, Matthew J.
   The slumbering masses : sleep, medicine, and modern American life / Matthew J. Wolf-Meyer.
   (A Quadrant book)
   Includes bibliographical references and index.
   ISBN 978-0-8166-7474-9 (hc : alk. paper)
   ISBN 978-0-8166-7475-6 (pb : alk. paper)
   1. Lifestyles—United States. 2. Sleepwalking—United States. 3. Sleep disorders—United States. I. Title.
   HQ2044.U6W65 2012
   362.196'84980973—dc23

                                      2012017393

Printed in the United States of America on acid-free paper

The University of Minnesota is an equal-opportunity educator and employer.

22 21 20 19 18 17 16   10 9 8 7 6 5 4 3 2 1

For my parents, Melanie and Robert,
who endured many years of my wayward sleep

CONTENTS

·················

ABBREVIATIONS

. . . .

AAP   American Academy of Pediatrics
ASPS   advanced sleep phase syndrome
BiPAP   bilevel positive airway pressure
CAP   Continuous Assisted Performance project
CAREI   Center for Applied Research and Educational Improvement
CPAP   continuous positive airway pressure
CPSC   U.S. Consumer Protection Safety Commission
DARPA   Defense Advanced Research Projects Agency
DSPS   delayed sleep phase syndrome
ES   excessive sleepiness
FFI   fatal familial insomnia
GHB   gamma hydroxybutyric acid
HCMC   Hennepin County Medical Center
KLS   Kleine-Levin syndrome
MCMC   Mississippi County Medical Center
MSDC   Midwest Sleep Disorders Clinic
NIH   National Institutes of Health
NREM   non-REM sleep
NSF   National Sleep Foundation
OSA   obstructive sleep apnea
RBD   rapid eye movement behavior disorder
REM   rapid eye movement
RLS   restless legs syndrome
SIDS   sudden infant death syndrome
SRED   sleep-related eating disorder
SWSD   shift work sleep disorder
TBYT   Take Back Your Time movement

PREFACE

···················

Sleep at the Turn of the Twenty-first Century

SIGNIFICANT CHANGES OCCURRED IN THE WAYS THAT AMERI-
cans conceived of and practiced sleep between 1996 and 2006. New
pharmaceuticals were produced and introduced to control sleep and
wakefulness, as well as a variety of other sleep complaints. New techniques were developed to confront daytime sleepiness at school and at
work, namely, the alteration of start and end times and provisions for
napping. New caffeinated beverages were introduced to appeal to youth
markets, and coffee consumption increased. Scientists worked to establish the genetic basis of early and late rising (so-called larks and owls)
and the causes of and cures for narcolepsy, restless legs syndrome, and
rapid eye movement behavior disorder. Alongside these changes, much
about American sleep stayed the same or intensified along familiar lines.
Insomnia continued to be a major complaint of many American adults,
with some estimates placing the rate of periodic insomnia near 30 percent
of Americans. Stimulant use to offset sleepiness and fatigue continued in
a fashion that stretches back to the early 1800s, albeit intensified. Popular
representations of normal sleep continued to focus on eight consolidated
hours of sound, motionless sleep, either solitary or in a shared bed with
an intimate. Clinical practice also followed these lines and focused on
establishing regular, consolidated nightly sleep for patients. Sleep was
emerging as a new interest but was shaped by very particular American
ideas about it and its ideal forms.

It was at this moment that I began the research that laid the basis for this book. What led me to become interested in sleep was a similar twin of emergent and historical forces. In 2002 I began developing a research project about the role of night in contemporary American society, thinking primarily about night work and the other productive efforts that are relegated to darkness to maintain our everyday lives. As part of that project, I imagined a chapter on sleep—that which occurs for the majority while these nocturnal labors ensued. I had begun thinking about this project because of my own experiences working third shift throughout college, which had attuned me to this often overlooked component of contemporary society. I worked four ten-hour nights, from 10 P.M. until 8 A.M., and fit the rest of my life in around that schedule, alternatively taking very early classes and night classes. I had been drawn to this work in part because of my tendency to go to sleep late throughout high school, only to struggle through the school day with a lack of sleep. But I was different from my coworkers, most of whom accepted third shift for the slightly increased pay that incentivized working an undesirable shift. For me, sleeping through the day posed no problem, and on days when I was off work I would maintain my nocturnal schedule; my coworkers had very different experiences. So when I found my way to the Midwest Sleep Disorders Clinic (MSDC) as a result of an article that had been published as a profile of the clinic and its researchers, I was primed to find sleep interesting.[1] At the time of my initial visit, I had no idea of the diversity of sleep disorders or of the emergent concerns of clinicians and researchers regarding sleep and the market forces compelling its insurgent interest in American society. After two hours at the MSDC, I had abandoned my earlier project, drawn to the research possibilities that focusing on sleep might allow.

From January 2003 through April 2007, I conducted archival and ethnographic research, first in the Twin Cities, in Minnesota, then in Chicago, Illinois. The research began at the MSDC, where I attended weekly case discussion meetings and departmental lunches as well as visited the overnight sleep clinic. I conducted formal and informal interviews with clinicians, researchers, patients, and their families at the clinic and throughout the Twin Cities. At the same time I attended local support groups for individuals diagnosed with obstructive sleep apnea and restless legs syndrome and conducted archival research at the Wangensteen Historical Library of Biology and Medicine, housed at the University of Minnesota and containing medical monographs from the nineteenth and twentieth centuries. In February 2006, I relocated to Chicago to conduct further research: archival research at the Special Collections Research Center, at the University of Chicago's Regenstein Library, and ethnographic research with patient support groups throughout the Chicago metro region. Coextensive with the in-depth field research during these two periods, I attended local and national professional meetings for sleep physicians and researchers and national support group meetings for less common sleep disorders, especially narcolepsy. I participated in online support groups for disordered sleepers, primarily Talk about Sleep, which allowed me access to individuals who might not seek out medical treatment or who might have intermittent complaints. Over three and a half years of research, I informally interviewed more than eighty disordered sleepers and conducted life history interviews with an additional forty. I attempted to interview an array of individuals who included diagnoses of all the currently recognized sleep disorders, but because of the rarity of some conditions this was not entirely possible. I conducted interviews with a dozen sleep clinicians and researchers, some affiliated with the MSDC. I interviewed sleep technicians, nurses, and family members of disordered sleepers. In the pages that follow, all but two of the interviewed individuals are rendered anonymous through the use of pseudonyms, but their descriptive information is otherwise unaltered. Only a handful of interviews are reproduced here; rather than produce "composite" characters to stand as representative examples, I have included interviews from a variety of disordered sleepers.

By all accounts, Americans have begun to sleep in a consolidated fashion (an average of eight continuous hours each night, sometimes referred to clinically as the "11 to 7 model") only in the past century or so. Prior to widespread industrialization and the eventual coordination of the workday in the United States in the nineteenth and twentieth centuries, most Americans probably slept in a less consolidated fashion, breaking their nightly sleep into two or more periods and supplementing this with daytime naps. Yet not all Americans today do sleep in a consolidated fashion, as current statistics regarding insomnia rates would seem to indicate and the narratives of disordered sleepers throughout this book support. But the 11 to 7 model of human sleep is appealing, one that motivates unconsolidated sleepers to seek medical treatment, physicians to care for nonnormative sleepers, and pharmaceutical companies to develop medications that produce consolidated sleep. The business in pharmaceuticals that produce sleep has always been steady, but in the mid-1990s new chemical agents were invented that produced sleep in new ways—the so-called Z-drugs (Sonata, Ambien, Lunesta, and Imovane). These new sleep-inducers have been appealing to patients and physicians since their introduction, and pharmaceutical companies have met consumer demands for drugs that prolong sleep through controlled-release technologies and higher dosages of basic chemicals. Simultaneously, drugs have begun to be marketed to other sleep disorders, including restless legs syndrome, narcolepsy, and "excessive daytime sleepiness." Technology has been developed for obstructive sleep apnea, and there is now a competitive market of manufacturers who sell various face masks, accessories, and decorations for their continuous positive airway pressure (CPAP) and bilevel positive airway pressure (BiPAP) machines. Although we have always slept, and there have always been those who have thought about sleep and its disorders, since the mid-1990s sleep has become big business, and American attention to sleep has intensified.

Homo sapiens, like all other species on Earth, have evolved over time to sleep. In the words of noted sleep researcher Allan Rechtschaffen, this may be "the greatest mistake evolution ever made" if it fails to "serve an absolutely vital function."[2] Whatever the reason why we sleep, the presence and rhythms of sleeping structure our everyday lives. Simultaneously, the structure of our everyday lives impacts our sleep. Our workdays, school commitments, family lives, the wakefulness of others, and our institutional obligations demand our presence and attention. As the institutions that make up our daily lives change—as the workday has lengthened, then shortened, then lengthened again—our sleep changes along with institutional commitments. The same might be said about all of our biological processes, including eating and defecating, the care of the old, young, and ill, our movement and respiration, our reproduction and aging. But our sleep is always with us, much like our appetite, yet when we sate our need for sleep we can hardly appreciate it. Instead, our lack or abundance of sleep is carried with us into the next day.

This book is about the social formations of sleep and the forms of sleep that society produces. Specifically, it is about how normative ideas about sleep have consolidated in the United States over the past two hundred years and how these ideas produce forms of life that are considered healthy or disordered. What is disordered sleep? Is there any way to know for certain? Is it not being able to sleep? Not getting enough sleep? Not sleeping at the right times? Not breathing while sleeping, or acting out one's dreams? Is an inability to control one's own sleep dangerous? For some of these disorders, there are clear, physiological causes; for others, the causes are not so simple. Sleep is always social, affecting others and affected by others. Society cannot exist without sleep, or sleeping without social expectations.

This perspective and these questions are indebted to my training as an anthropologist; I approach these concerns about sleep with a background in ethnographic research, a practice concerned with representing the many stakeholders and contexts that undergird and make lively the contemporary world of sleep, including scientists, physicians, disordered sleepers and their bed partners, and activists and social scientists interested in health and public policy. I have a background in literary analysis and historiography, which leads me beyond medicine and science to the popular and historical contexts of American sleep; throughout this book I supplement interviews and ethnographic descriptions with textual analysis and historical detail in an effort to elucidate both the contemporary moment of American sleep and its broad purchase in our everyday lives, as well as the many historical debts our modern conceptions of sleep owe to the particular contexts in which the science, medicine, and culture of sleep have developed. Like many anthropologists of medicine and science, I am not a trained physician or laboratory specialist but rather a sympathetic critic who received training in the fields of my interests—in this case neuroscience and the science and medicine of sleep. In some cases, I am critical of the emergent or historical concerns of sleep science and medicine, and this criticism originates in the holistic perspective granted by anthropological training. At other times, when the science and medicine under discussion are robust and well-founded, I consider their ramifications on American experiences and expectations of sleep and everyday life.

As much as this book is about sleep, it also moves beyond sleep. Examining sleep provides ways of thinking about American society over the past two centuries, its development of labor practices and markets, the rise of allopathic medicine and its treatments, and our science fictional, juridical, and military fantasies of human physiology. Focusing on sleep changes the way we understand our intimate, chemical, and biological selves and relationships with others. Close scrutiny of the control of sleep on the part of institutions and individuals exposes how American everyday life is formed through spatial and temporal ordering of practices and events. These are the foci that guide this book. They lead in overlapping and tangential directions, but at their base is the question of sleep—how it shapes our lives and how it makes forms of life that are recognizable as healthy or disordered.

INTRODUCTION

· · · · · · · · · · · · · · · · ·

From the Lone Sleeper to the
Slumbering Masses

Why do we sleep the way we do? Thomas Edison, who was a titan of industry, shaping the twentieth century through his inventions, business practices, and beliefs, provides one answer to this question. Among his many claims to fame are his views on sleep, a product of everyday life at the turn of the twentieth century:

> Most people overeat 100 percent, and oversleep 100 percent, because they like it. That extra 100 percent makes them unhealthy and inefficient. The person who sleeps eight or ten hours a night is never fully asleep and never fully awake—they have only different degrees of doze through the twenty-four hours. . . . For myself I never found need of more than four or five hours' sleep in the twenty-four. I never dream. It's real sleep. When by chance I have taken more I wake dull and indolent. We are always hearing people talk about "loss of sleep" as a calamity. They better call it loss of time, vitality and opportunities.[1]

Although not as well known as his contributions to carbon filament electric lights, the phonograph, and telegraph enhancements, Edison's views on sleep are pervasive and particularly American. You might not know them word for word, but you know the spirit of them: too much sleep is a waste of time; too much sleep is an indulgence. It might be okay for some people at some times, especially on a weekend morning, but every

1

hour slept through is an hour not spent at work, or school, or play. Every hour spent sleeping is an hour that could have been spent doing *anything* else. And Edison is hardly alone in his views.

But Edison was a paradoxical sleeper. Although he might not have slept more than a few hours at night (four or five), he napped throughout the day. He had cots available to him wherever he worked, and he would doze from time to time, supplementing his nightly sleep with naps. There is no evidence as to how much sleep Edison actually got throughout the twenty-four-hour day, but four or five hours of sleep each night, followed by one or two naps the following day add up to not a particularly exceptional or heroic amount of time spent asleep. Add to this the apocryphal story of Edison's use of naps to spur creative thought. Sitting in a chair resting on a wooden floor, Edison would hold a metal ball bearing in his hand. When he fell asleep, his hand would relax and he would drop the ball, awakening him. From this twilight state, just beginning to fall asleep, Edison claimed he was able to solve problems that vexed him when conscious. How often he did this is unclear. But despite Edison's scant time asleep at night, he was hardly incessantly awake either, especially considering his naps during the daytime. However, Edison is remembered as a short sleeper and an ingenious one at that. It's this Edison whom Americans choose to remember, and his sleeping provides a basis for thinking about ideal sleep.

Dominant conceptions of sleep have changed over time, and so have American understandings of the normal, the ideal, the disorderly, and pathological.[2] These changes are the result of scientific, medical, and popular forces: the identification of circadian rhythms that govern feelings of alertness and sleepiness, the nosological definition of various pathological forms of sleep, and commercial and media representations of sleep have all collaborated in altering how sleep is conceived. These are all changes that have occurred since the 1880s and been dependent upon social formations and cultural expectations from early colonial America through the Industrial Revolution. How Americans sleep now is tied to these historical forces and human physiology. The biological is indebted to the social, and the social is inextricably bound to the biological. Neither is primary in its status and power, nor is either subservient to the other. And sleep is an exquisite example of the intermingled forces and effects of such intimate social and biological formations.

Americans have not always slept the way they do now. By all accounts, Americans, like other people around the world, used to sleep in an unconsolidated fashion, that is, in two or more periods throughout a twenty-four-hour period. Sometimes referred to as "first" and "second" slumber, nap, or sleep, a day's sleep was divided between the night and the day.[3] With the advent of the consolidated workday in the 1840s and onward, while struggles were waged over how long that day would be, ranging from twelve to eight hours, sleep was also consolidated.[4] At least this was the case for many Americans. But there have always been those who sleep otherwise: night workers who sleep through the day, parents who nap while their children do the same, rebellious sleepers of all sorts, as well as all those who sleep in a more disorderly fashion: insomniacs, narcoleptics, and their kin. And in much of the world, because of a lack of electric light and unconsolidated workdays, people still sleep in an unconsolidated fashion. There is nothing particularly in tune with nature, or primal, or healthy about unconsolidated sleep: it, too, is subject to social formations and cultural expectations, albeit of sorts different from those associated with consolidated sleep. The most natural sleep may be sleeping whenever one feels tired and being awake whenever wakefulness takes hold. But who is so fortunate to sleep so spontaneously?

As a result of the ever-present biological and social impacts on sleep, unbiased conceptions of sleep are not possible. Science can never uncover the truth about human sleep, because sleep is always biological and social, cultural and natural, historical and emergent, and will always be perceived through contextual lenses.[5] Ultimately, no certainty can be had. This undergirds the science and medicine of sleep with a foundational doubt: What is sleep, and how can it be known? Instead of being able to answer these questions definitively, sleep researchers have defined models, averages, norms, and pathologies, providing the basis for medical intervention.[6] These conceptions of normalcy compound and are compounded by everyday spatiotemporal rhythms that shape the lives of individuals and institutions;[7] sleep binds individuals to institutions, and when disordered sleep disrupts these interactions, medicine intervenes to reorder the everyday. These everyday orders, in turn, structure American capitalism, a form of capitalism that is tied to long-standing conceptions of normalcy, medicine, and everyday life. If this sounds vertiginous, it is in no small part due to the historical and contemporary interactions of medicine, science, and capitalism that formulate spatiotemporal orders of everyday life that in turn produce our contemporary desires for sleep. These desires bind individuals to society in intimate ways that are complex and difficult to isolate or clarify causally.

Sleep is not something that anthropologists have focused on over the history of their discipline, and my interest in the topic is indebted to my theoretical concerns with everyday life and its spatiotemporal ordering. My intellectual influences are uncommon for anthropologists, especially within medical anthropology. Critical medical anthropology, which distinguished itself from more positivist medical anthropology in the 1980s, has been marked with an interest in the application of the work of Karl Marx and Michel Foucault. From Marx, medical anthropologists have focused on the complex political economies of health, illness, and social suffering, working at once to expose how material conditions produce forms of life and how those conditions that are pathological are indebted to economic interests on the part of corporations, the elite, and the state.[8] From Foucault, medical anthropologists developed an interest in power and subjectivity, particularly in how institutions produce individuals and how individuals come to recognize themselves as individuals. Foucault's influence has been especially apparent among feminist medical anthropologists, who have found robust descriptive language in Foucault's work to critique medical institutions and their production of normative ideas of life.[9] More recently, medical anthropologists have found inspiration in the work of science studies scholars, especially that of Bruno Latour, who, in his descriptive mode, has helped to refocus attention on the agentive capacities of human and nonhuman actors despite their perceived powers or lack thereof. Science studies scholars have also opened up the terrain of analysis, moving beyond the clinic and the community to conceptualize the roles of laboratories, trade networks, and other tertiary sites as vital to the interactions of medicine, physicians, and patients.[10]

All of these approaches are important to my own work, and I build upon them with other theoretical concerns that stem from four other French thinkers: Henri Lefebvre, Bernard Stiegler, Gilles Deleuze, and Félix Guattari, who raise questions of immanent spatiotemporal formations and their interactions with bodies. What brings these four thinkers together is their collective interest in the everyday, although both Deleuze and Stiegler refrain from rendering their philosophical thought in that vein. For Lefebvre, what binds the diverse and sometimes contradictory events of our lives together is the concept of the everyday.[11] The everyday designates both the realm of the ordinary and the spatiotemporal rhythm of things, marking what is commonplace as well as the contexts in which it is unremarkable. Moreover, the everyday cuts between temporal frames, aligning the present with the past and the future. The everyday is such because it occurs *every* day, albeit with some variation. As such, the everyday also binds spatiotemporal orders across scales, from the day through the week to the month, season, and year, allowing for variation within expectations.[12] This question of difference and repetition is central to Deleuze's thought, both his individual philosophical works and his collaborative projects with Félix Guattari. This is especially apparent in Deleuze's discussion of two interrelated concerns, habit and "bodies without organs."[13] For Deleuze, the individual is formulated through a complex interaction of internal drives and external forces, which, combined, bring into being the desires that direct us through life. Our "organs" lie outside our bodies, and it is only through their arrangement that we are recognizable as individuals to ourselves and others; we are our desires, which bring our individual lives into contact with external forces. Habits are both the basis and the object of desire; they make up our everyday activities. However, because of their repetition, habits provide the possibility of unexpected or uncommon outcomes; through their very repetition, they open up the possibility of difference, thereby opening the past and present into an uncertain future. This question of the future, and one that is always doubtful in its capacities, is the foundation upon which Stiegler has elaborated his concerns with technics, with the fundamental relationship of humans and technology.[14] Inverting the schema of historical materialism, wherein humans are indebted to the material conditions of life in an always past-focused orientation, the potentiality of technology—its possibility for different futures—is the basis of Stiegler's understanding of technicity. Technology has embedded in it both specific histories of development and particular futures that it can facilitate. In our interactions with the world, we produce emergent futures, everyday lives indebted to the past but always unfolding in unexpected ways and open to new possibilities.

This interest in the emergent highlights the processual quality of being in the world.[15] This orientation stresses the always changing, always becoming nature of objects and subjects, individuals and communities, environments and subjectivities; everything is in the process of formation. No occurrence of sleep is an exact replica of an earlier event. Every time we sleep, we repetitively enact similar practices (brushing our teeth, putting on pajamas, relaxing by reading), but minute differences set every sleep event apart from others. Attending to the always changing nature of sleep and everyday life destabilizes them as inert, explanatory objects and instead renders them complex processes binding past, present, and future in irreducible ways. This emphasis on the processual, on the emergent, moves beyond the sometimes static concerns with political economy, disciplinary institutions, and actor networks that often stress historical and contemporary conditions that render processes as objects.[16] Moreover, and more important, recent interests in expertise, pathologization, and imagination become inverted and rendered as doubt, desire, and thresholds; they become prognosticatory, bringing the future into the present, and are thereby capable of formulating new futures.[17]

Doubt in Medicine

At its base, medicine is about doubt and certainty. The process of interpreting a set of symptoms and arriving at a diagnosis imposes a claim of certainty to allay doubt that the condition might be interpreted otherwise. But these diagnoses, these claims of certainty, can sometimes be undone, by the presence of emergent symptoms, the ineffectiveness of treatment, or the misinterpretation of the symptoms. Take, for example, the clinical use of the term *idiopathic,* meaning "cause unknown." Symptoms can be recognized and diagnoses provided, but this process can be uncoupled from knowing what the cause of the symptoms or disorder is. Cause can often be disassociated from cure as well. Since most treatments focus their powers on symptoms, the reason for someone's insomnia can be inconsequential to its treatment, if that treatment is effective. Because an effective treatment can provide the cessation of symptoms, it can obscure the underlying cause of a disorder. This can result in adverse reactions to treatments; sometimes the treatment of one disorder can produce new, secondary disorders. This unfolding of symptoms and disorders belies medicine's claims to certainty; the possibility of doubt sustains and motivates medical and scientific practice.

Sleep medicine, especially, is infused with doubt. There are lingering doubts about what sleep is, as there have been since the inception of sleep science and medicine. Why do we sleep at all? What function does it serve? How much sleep does an individual need? How much sleep is healthy or pathological? Why are some people early or late risers, larks or owls? Can you be partly asleep and partly conscious of your actions? Can we be rid of the need to sleep altogether? To a certain degree, these kinds of questions might be asked about any number of physiological processes, from eating and excretion to reproduction and respiration, but sleep stands out among our basic biological processes as being subject to an intense questioning about its fundamental functions and variations.

The etiologies of many medical disorders similarly defy explanation—is it genetics, the environment, lifestyle choices, or random accident that leads to autism, cancer, muscular dystrophy, and cerebral embolisms?—but these disorders are not constituent of life itself, as sleep is. We might ask what autism is, but that question is predicated on knowing what normal development and behavior are in order to map the variance of the disorder. There are scientific and clinical expectations of what normal sleep might be, namely, eight consecutive hours of nightly sleep. But this is an average, and clinicians acknowledge that anything between seven and nine hours of nightly sleep can be normal as well. In addition, as sleep requirements change throughout the life course, what is an appropriate amount of sleep at one age is insufficient or too much at another point in life. Added to this are the complexities of many of the sleep disorders: Is this bout of insomnia a momentary, stress-induced case, or is it based in a neurological or environmental cause? Is this a case of narcolepsy or consistent sleep deprivation and excessive sleepiness during the day? Clinicians are faced with these questions on a daily basis, and how they answer them has as much to do with how they understand sleep and its variations as it does with the cultural milieu they operate in, which demands the application of their expertise to propose a solution to a patient's disordered sleep.

The process of isolating a sleep disorder from its environmental context—often including the very stuff of American everyday life: food and drink consumption, family, school and work obligations, media diversions—requires discerning the influences of time and space on individuals and their lives. American everyday life is spatiotemporally complex, particularly because of its commonsensical nature: When did you wake up this morning? How long did you sleep last night? Was it restful? Do you remember being woken up? How about your drinking and eating today: How much sugar and caffeine did you consume? When did you eat, what did you eat, and how much of it? How much light exposure are you getting? How much exercise and at what times of day? Do you share a bed? With whom, and how does he or she sleep? How do your sleep and daily commitments change over the course of the week? All of these factors can have profound influences on sleep but are often taken for granted. Patients can be asked to take vacations or sick leave from work, catalogue daily activities and food and drink consumption, and refrain from negative influences on sleep in an effort to let individuals "free run" with their disorders. Free running attempts to isolate the disorder from its environmental context and can often show how individuals or their environments are the cause of or the exciting influence in their disordered sleep. The therapeutic result of this process often focuses on the control of an individual's body and its relation to the spatiotemporal expectations that frame an individual's everyday life. But for many patients, undergoing this process is impossible because of social obligations, and the cause of their disordered sleeping may never be known.

Medicine and Spatiotemporal Control

One method to allay these doubts is the formulation of normative spatiotemporal regimes. This occurs at the individual, subjective level and at that of institutions, populations, and society itself. In the case of sleep, this can be seen in the eight hours of consolidated sleep that individuals seek, which structures institutional uses of time and is the basis for pharmaceutical treatments. Each reinforces the others, thereby producing a robust cultural logic of sleep, spatiotemporal rhythm, and normalcy. When an individual has the experience of disordered sleep, when his or her desire for sleep fails to align itself with the spatiotemporal orders of everyday life, medication offers one means to alleviate this disorderly desire for sleep.

This medical control of time stems from a long history of management of populations and their uses of space. Two cases in which this is most apparent are the intensification of urban spaces and the colonial governance of populations. Both are predicated on the colonization of bodies through medical management and control. Outbreaks of epidemic disease make this evident: the cholera outbreaks in the United States and Europe in the nineteenth century necessitated broad governmental action based upon extant medical knowledge of transmissibility and susceptible populations.[18] Similarly, in the British colonies of India and the American colonies in the Philippines, cholera, bubonic plague, and leprosy provided the basis for the spatial management of populations, resulting in quarantine and leprosy colonies.[19] More recently, outbreaks of SARS and avian flu have made evident how recurrent these techniques are.[20] In each of these cases, extant medical knowledge has carried with it cultural expectations of health and illness, often reinforcing social divisions along ethnic and class lines. In the case of cholera in the United States, the poor were seen as dirtier and more susceptible to disease, whereas the bourgeois class was popularly accepted as being largely protected from cholera. But the disease showed no respect for class divisions, and the wealthy and the poor alike succumbed to cholera. In India and the Philippines, medical knowledge provided the basis for colonial control of populations, further entrenching the divisions between those in power and their colonial subjects. Eventually the techniques that were used to manage colonial populations came to be used on populations in imperial metropoles, the working class and the poor. Because medical knowledge is always also cultural knowledge, it carries the burdens of its contexts, often reproducing expectations of difference. Medical professionals conceive new orderings to the world and thereby produce forms of life that are healthy and unhealthy, orderly and disorderly, but are always indebted to the cultural situations in which medical knowledge is produced and applied.

The control of space is also the control of time. Quarantines come with expectations of their termination, medical understanding of a disease's development and treatment, the need for policing and care, and the timing of medical interventions. Hospitals, similarly, confine patients to particular spaces, allowing periods of movement and visitation and also providing medical professionals with reliable access to patients for treatments. Many medications have temporal requirements, either limiting the amount of a drug to be taken or requiring regular consumption of a prescribed medication. Requirements to take drugs every set number of hours or twice per night, as in the narcolepsy treatment Xyrem, necessitate that individuals coordinate their proximity and access to drugs. To take the example of Xyrem, narcoleptics require two doses each night to consolidate their sleep. They take the first dose when they lie down to sleep, and after four hours of sleep, they awake to take a second dose. By doing so, they sleep soundly for the four hours that the dose's effect lasts. Although Xyrem is a strong case, it makes evident how both proximal and timed access to medication structures the lives of individuals; being far from the drug when sleep is desired means either relocating or stalling sleep, which may have consequences. Other drugs, other modes of treatment, similarly require temporal and spatial coordination, thereby formulating everyday orders and disorders.

Beyond Xyrem, how is this control of space and time manifested in sleep and its therapeutics? At its base, all sleep medicine is about ordering the lives of individuals to reliable and rhythmic spatiotemporal expectations: eight hours of sound, restful sleep each night and daytime alertness with a minimal sense of sleepiness or fatigue. For individuals who struggle with obstructive sleep apnea symptoms, ready access to their CPAP or BiPAP machine is necessary for a night's sound sleep. For the nightly alleviation of symptoms associated with restless legs syndrome, REM behavior disorder, insomnia, and circadian disorders, regular medication is required. In the case of many disorders that require only a pharmaceutical treatment, medical intervention has been rendered portable. In this portability, individuals can produce spatiotemporal effects wherever their medication allows, whether at home or work or elsewhere. The powers of shaping the temporal and spatial experiences of individuals have moved out from the hospital and quarantine to be embodied in everyday prosthetic, pharmaceutical, and behavioral treatments. This therapeutic milieu, in which the means of medical intervention are portable and rely upon the actions of individuals (their "compliance"), typifies contemporary American medicine, which is especially evident in the treatments for sleep disorders. With a pill or two or a machine, disorderly sleepers can produce sleep that meets their spatiotemporal expectations. The control of sleep is within the means of individuals, given that they have access to the necessary medical treatments.

These expectations are built upon normative ideals of where and when sleep takes place. Why do we sleep in rooms set aside for this sole purpose? Why sleep in beds as we do? And why the social arrangements of sleeping that we have: no more than two to a bed? Take, for example, Select Comfort's Sleep Number bed, which is billed as "the perfect bed for couples."[21] Central to this claim is that individual sleepers have distinct needs for comfort and that one bed needs to serve two quite different bodies. The Sleep Number bed is composed of two air-filled mattresses that can be modified in their level of support through a small remote control on each side of the bed. The sleeper can set the desired density, which increases in increments of 5, from 0 to 100, allowing one side of the bed to be different from the other. Two air pumps located under the bed regulate the pressure in the mattresses and periodically reinflate the mattress to the desired level. The logic behind the Sleep Number bed is rather straightforward: the higher one's body mass index, the lower the recommended density of the mattress. Because the average distribution between adult bodies is rather narrow, so too are the options available in the Sleep Number bed. But this is all beside the point: Sleep Number beds trade on the expectation of Americans that male and female bodies are necessarily different and yet are to be brought together on a nightly basis for bed sharing. Moreover, a bed is for two people and should be soft. But, first and foremost, one's nightly sleep is about the experiences of an individual body. In a recent Sleep Number television commercial, a woman reports, "I actually enjoy sleeping next to my husband now." Presumably sleep before acquiring a Sleep Number bed was less intimate or intimate in a less desirable way. Therapeutic technologies assist individuals in making their disorderly life into something decidedly more normal, providing them with the ability to control their intimate interactions.

The ability to control disorder is a dual-faced situation. At once it allows individuals to have power over their symptoms, thereby granting a sense of efficacy. However, this is paired with responsibility and obligation, what is accepted as "compliance" in contemporary medicine.[22] Being able to comply with one's treatment is a complex burden. Beyond access to treatments, and effective treatments at that, individuals also need economic and social support for the use of the treatment. This requires holding a steady job for access to medical insurance and wages or state-supported medical care. It also requires the care and interest of family members, who will be exposed to the side effects and the financial burden of treatments. In the disorderly sleep of individuals, family members and intimates are enrolled, as are the institutions that make up one's everyday life. The disorderly sleep of individuals reaches out to society, to those masses of other sleepers who provide the basis for understanding one's sleep as normal or abnormal.

Formations of American Everyday Life

Formations bring together and simultaneously isolate bodies, actions and practices, rules and expectations, acts and statements, ideologies, and understandings into an orderly world; formations are irresolvably material and abstract.[23] Formations are both historical and unfolding, reproduced and changing through everyday activity. Our everyday lives are formations that bring together spatiotemporal rhythms, individuals and institutions, and order and disorder. As such, formations depend upon institutions; institutions are the forms that produce variable expressions of the content implied in the formation.[24] Work, school, the family, the legal, the medical, among other things, provide this foundation for formations of everyday life. And the very idea of everyday life binds these diverse activities and institutions into a whole.[25] One person's work life is necessarily different from another's, as is work life from family life. But these institutional domains provide the fabric with which we weave our lives, and our lives become recognizable to ourselves and to others through these associations and investments. But our everyday lives fray and come undone through their reproduction, leading to their reformation. Each day is different from the last but subject to expectations of what the day should be. The day and its rhythms shape our lives and are shaped by the forms of life that they produce, especially orderly and disorderly sleep.

Central to our experiences of everyday life are its spatiotemporal rhythms. When do we wake and sleep? Eat, drink, and defecate? Work, attend school, and participate in family life and recreation? Celebrate holidays and vacations or meet deadlines imposed by institutions and seasonal shifts? From the micropractices of our everyday lives (making a cup of coffee, showering and dressing, eating meals, traveling, working, and attending school), we are intimately connected to the everyday rhythms of others. These spatiotemporal rhythms guide us in our individual daily lives but also order society.[26] Our rhythms are intimate ones—we know when we desire a meal, a drink, sleep—and part of that intimacy is their regularity. Our regularities are those of others, and this shared coordination, often implicitly produced, binds individuals into social formations. Sleep is an exquisite example of a spatiotemporal social formation that is based upon regularity. Clinicians refer to it as the "11 to 7" model of sleep, meaning that individuals sleep from about 11 P.M. until 7 A.M. the following morning, when they wake up to prepare for work. What might feel an intimate part of one's everyday life, namely, beginning to experience sleepiness close to 11 P.M., preparing for bed, and eventually turning off one's bedstand light, are actions that are shared by millions of Americans in roughly the same way and at the same time. This great aggregate of regularity, that millions of Americans retire to bed at the same time and wake at approximately the same time, structures such diverse things as television and radio programming, traffic signals, the opening and closing of businesses, police patrols, and so on. From the simple and coordinated act of retiring to bed, we might extrapolate to all of the spatiotemporal rhythms of American society and their connections to individuals. These reliable spatiotemporal rhythms depend upon expectations of normalcy that at once structure the lives of individuals and the everyday orders of society.

Since the middle of the nineteenth century, conceptions of normalcy have provided a basis for the practice of medicine and social formations. The idea of the norm arose with the ability to produce and use statistical averages.[27] This, in turn, depended on a widespread interest in the normal, primarily for governmental use and particularly in relation to industrial pressures as well as the possibilities of experimental science. Some have argued that prior to the norm, ideals provided individuals with aspirational models: one might aspire to be like the heroes or maidens of myth, but one could never actually become like his or her ideal.[28] The invention of the norm has been accepted as a transition away from ideals, a productive move, in the sense of productive forms of power,[29] in that everyone might become normal, thereby providing a trajectory for his or her desire. Ideals are prohibitive in that individuals can never achieve them; norms, on the other hand, are accepted as such because they can be achieved. Added to this, the formation of norms has been coextensive with the development of the category of disability, for norms imply variance (everyone is judged by his or her distance from the norm, exceptional either in over- or underachievement), whereas ideals are operative and all individuals fail to achieve them. Following from this, everyone is slightly disabled from the perspective of the norm: even above-average intelligence or good looks are "abnormal." And those people who severely differ from the norm are categorically "disabled." The formations and popularizations of norms *and* ideals provide parallel lines of desire, because both depend upon a foundation from which variance can be articulated.[30] In American society, the gross semiotics of the body provide this point of divergence and articulation of the self, as do discourses of health, although ideas of the norm are diffuse and inflected by subjective experience of what the norm should be.

How do we move from the problems of individuals to the problems of society? How do we move from the individual to the masses? Central to the theorization of pathology and normalcy is the individual, person, and self. These isolated concepts (or concepts of isolation, that a part can be removed from the whole) find their logic in their similarity to or difference from conceptions of populations. Populations are in turn produced through recourse to numbers derived from individuals. This interaction between part and whole is predicated on who counts, who can be aggregated into a population as a representative individual. The assumption of belonging, of mattering, is fraught with formations of society that include some but not others. Integralism relies on marking some individuals as mattering, as well as designating those who do not.[31] In this process, integral formations also generate mechanisms for being recognized as mattering, for coming to be recognized as part of the masses. For example, consider what is usually referred to as "medicalization": what was once considered natural, as in the case of periodic and self-resolving insomnia, is now treated as a medical disorder.[32] Where once you might have dealt with short-term insomnia with a warm drink before bed and a relaxing bath, now pharmaceuticals are the taken-for-granted treatment for even minor insomnia complaints, and individuals come to believe in the necessity of medical intervention. This is the result of the power of contemporary medicine in American society, which serves as a mechanism for aligning individuals with the institutions that make up their everyday lives through the proliferation of desire. For those who are outside the masses, outside the normal, medical treatment is one means to render them part of the masses, integral to society. As a result, individuals appeal increasingly to the powers of medicine, not solely for the alleviation of symptoms, but as a form of citizenship, a form of belonging, mediated by a desire for normalcy.[33]

Desire mediates relationships among individuals and between individuals and institutions. In this process of mediation, desire produces discursive realms: the biological, the social, the cultural, the economic, the chemical, the natural.[34] That we take these to be different and at times divorced from one another, that we imagine the biological can ever be separated from the social or the cultural, is an effect of purification.[35] The reification of discursive realms lays the foundation for the proliferation of disciplines of expertise and also serves as a means to isolate the causes and effects of disorder, ultimately resulting in the reification of the biological, social, cultural, economic, chemical, and so on.[36] That physicians are the experts to tend to sleep disorders has to do with the assumption that biological causes require medical treatments rather than an attempt to reorganize institutions to allow for variations of human sleep. And this is indebted to the historical forces that shape desire and its manifestations. Thus, desire is also always historical and future-oriented.[37] It is formed through the constant and modulating interactions between individuals and their worlds, which are based upon historically entrenched forms of life. But our desires pull us ever forward into repetitive and always different interactions, allowing for the possibility of emergent forms of desire and life.[38] These historical debts and future unfoldings are often capitalized upon in the form of primordial thought: fictional pasts are posited as other ways of desiring in the world, and these lead to new forms of desire and life. This can be seen in recurrent claims to a balanced, harmonious, agrarian past for Americans, when sleep was more natural.[39] But to assume that there was ever a time when sleep was more or less natural is also an effect of purification, an assumption that we have fallen from grace with nature, and the only way to restore it is through an eradication or modification of our modern present.

Desire is about both whole bodies and their parts; it is about bodies and their interactions with other bodies but also interactions internal to individual bodies. My desire for stimulation, my appetite for caffeine, has as much to do with deep-seated chemical dependencies as with my want of a particular somatic feeling, my taste for a particular chemical composition, and my need to be able to attend to my daily responsibilities. To support my modest and legal addiction, I am caught in expectations of labor and productivity, global networks of trade, the weather patterns of faraway places, and the lives of coffee bean famers. Insomniacs are tied in similar fashion to pharmaceutical companies, their research and production facilities, distribution networks, the Food and Drug Administration; the prescription and daily intake of a drug are never solely about an individual and his or her chemical treatments. In perceptible and imperceptible ways, desire binds forces together into forms of life that obscure their interconnections. Thus, desire is collective and individual. It is about the masses but also about individuals and their subjective relations to those masses. Desire binds parts to wholes, individuals to masses; it is connective and metonymic, collapsing scales into the rhythms of everyday life. Desire for sleep is always individual and collective; it is about bodies and their processes; it is about histories and futures; it is always social, biological, natural, cultural, economic, and chemical. Desire for sleep is irreducible, manifold, and influenced by the worlds we inhabit and those we create. But desire is shaped by its formations. It is neither infinite nor unfettered. Our desire for sleep is molded by the institutions that make up our everyday lives: work, school, family life. Desire for sleep is weighed against desire for labor, for stimulation, for interaction and intimacy. Desire is always multiple and generative of new formations: it unfolds into the future, proliferating emergent forms of life. And desire is always entangled with capitalism, which shapes and legitimates the desires and intimacies of individuals and their everyday lives.

A Somnolent Capitalism

Social science literature in the latter half of the twentieth century often accepted the continued rise and power of the capitalist market as an inevitable conclusion of globalization and market integration.[40] Added to this are the concerns emblematized in the idea of medicalization: what was once uncommercialized could now be marketed to a broader and broader population, given appropriate contextualization and incentive. This may seem naïve or triumphalist, but the conditions of the market economy in the United States and the wider world helped to ensure this popular vision. And it was a vision that translated beyond spatial imaginaries—as globalization does and the colonization of new populations does as well—to temporal ones, as in the case of Murray Melbin's characterization of "incessant" institutions.[41] Looking forward from the late 1970s, Melbin forecasted a United States, if not an entire world, that operated in a twenty-four-hour fashion, in which the primary difference between those who work through the day and those who work through the night is one of choice. In fact, the 1980s and 1990s looked as if they might follow Melbin's predictions, as grocery stores, gas stations, and restaurants lengthened hours and in some cases remained open for twenty-four-hour periods throughout the week or on specified days (usually the weekend). But technology changed this: the rise of the Internet provided people with easy, twenty-four-hour, at-home shopping. Credit card readers on gas pumps allowed gas stations to be untended while still serving needy drivers. Restaurants and grocery stores confronted the need to pay employees for third-shift work (usually twice what a day worker commands), and the meager profits garnered overnight failed to offset these costs. The incessant drive toward staying open late or for twenty-four-hours, which I will return to in Part II, met the intractable expectations of employees and customers and was found lacking. Or simply improbable. Why go to the movies in the middle of the night when video games, DVDs, and the Internet ensure that leaving the house is an option, not a necessity? So what began as the "colonization of night," in Melbin's words, resulted in the formation of another everyday logic of American social life, one that centers on the home and individual lives rather than on widespread social transformation.

If social formations were to be incessant, one of their logical outcomes would be incessant forms of life: the need to always be alert, never to sleep or to sleep in a very limited fashion. Instead what we see are forms of intensification: we are meant to be alert and attentive when awake and deeply asleep when sleeping. Rather than insomnia providing a model for an incessant form of life—one in which we are always awake, alert, and productive[42]—narcolepsy has become the rule. Narcoleptics require medication both to remain alert throughout the day and to sleep soundly through the night. Although they offer a strong case of this sort of chemical control of consciousness and their lives might seem a far cry from the way that many individuals live, the expansion of caffeinated beverages, legal stimulants, and napping strategies indexes a growing desire for alertness. Similarly, the rise in both sleep-specific medications (Ambien, Lunesta, Rozerem) and over-the-counter drugs that include sleep-promoting chemicals (such as Tylenol PM) marks a shift in pharmaceutical consumption that is focused on a desire for consolidated sleep. Added to this are new mattress technologies, like the Sleep Number bed, that promote the individualization of sleep. Although we may want to have more incessant forms of life—something on the model of Edison's ideal arrangement of sleep, if not perpetual wakefulness—what we have instead is an intensification of sleep and wakefulness, a model that accords with the ways that individuals, institutions, and everyday worlds are conceived early in the twenty-first century.

Capitalism has always been wrapped up with understandings of the human body, its capacities and potentials. As early as 1844, Karl Marx had diagnosed this function of capitalism, referring to it as the simultaneous production of alienation and "second nature," as laborers are divorced from the world in which they labor through the production of "value" but increasingly understand their relation to nature through this very labor.[43] What has changed, in part through technological advances that allow finer and finer understandings of the human body, are the scales at which capitalism is tied to the body. If in the nineteenth century, capitalism was largely focused on the whole body of the worker and its potentials for labor, the twentieth century saw the advent of molecular conceptions of the body, which provided new levels of control and profit.[44] By way of example, one might consider Frederick Taylor's theories of "scientific management," which take as their object the whole body and its functioning.[45] To improve efficiency, Taylor measured how long particular actions take (e.g., how long it takes a fit man to fill a wheelbarrow with coal) and extrapolated those numbers to whole workdays to establish benchmarks for productivity. This laid the basis for Fordist models of the assembly line, which coordinate bodies through spatiotemporal understandings of their capacity for productive labor. But how might bodies be made more productive? How might they sustain their attention longer? The colonial trade of sugar, coffee, and tea, according to Sidney Mintz, stimulated productivity and helped workers consolidate their sleep by working through the day. The introduction

of contemporary energy drinks, sleep- and alertness-promoting pharmaceuticals, and caffeine to the workforce, substances now integral to
American everyday life, is one way in which bodies and their capacities
are seen as being influenced by molecular forces. These basic chemicals
are the descendants of this capitalist interest in the shaping and control
of human biology. And, at the heart of these conceptions of human biology and its potentials is the need for reproduction, the need for reliable,
rhythmic order. This ordering is always bound to cultural expectations
of normal human bodies, their limits and potentials, and orderly society.

I raise these qualities of capitalist formations in an effort to make
local forms of capitalism more distinct from one another. American expectations of time and bodies are different from those of other societies. Rather than accepting capitalism as an abstract and universal system
of thought and practice, we should conceptualize it as an assemblage of
forces, exposing how capitalism is predicated on cultural expectations
of normalcy and order. Capitalism is not *a* system that appears the same
everywhere, nor is it *a* thing in itself. Capitalism is an assemblage of many
parts; how they fit together determines both local iterations of capitalism
and the parts of which it is assembled. Take, for example, the turn toward
"flexible accumulation," which many scholars took to typify advanced
global capitalism in the late 1980s through the first decade of the 2000s.[46]
The need for flexibility in terms of accumulation strategies, which involve innovative modes of production and distribution, as well as intensified models of consumption, also requires individuals and institutions to
adopt flexible strategies. From the 1970s through the first decade of the
2000s, workdays stretched beyond eight hours; workweeks expanded
beyond Fridays; employees required retraining and adopted the need to be
"flexible" themselves, adapting to ever new demands in the workplace.[47]
This has not been the approach adopted by all "capitalist" societies around
the world, but rather by those in which discourses of "flexibility" have
resonated, particularly the United States.

And here Edison offers an example: at a time when the management
of whole populations was necessary for the smooth and orderly advance-
ment of industrial production and exchange, an ordering of bodies that
led to the development of sleep science and medicine in the nineteenth
century, Edison took it upon himself to shape his sleep to his desires. But
he takes his desire for sleep and alertness to be worthy of a rule for all

Americans, if not all people. His ability to sleep as he did was particular, but the expectations that he embodies stand in for American sleep more generally: although we sleep as part of the masses, each of us addresses our sleep individually. We have specific desires for sleep and alertness, we know ourselves through our preferences for sleep, but our sleep is more similar to that of our fellow sleepers than it is different. Edison's interest in managing his sleep in the way that he did, his assertion that it could be successful for all Americans, if not all sleepers, is an outgrowth of modernizing attitudes toward sleep. Edison's attitudes index not only his care of himself, his recognition of himself as a sleeper, and his capacities for sleep and alertness, but also qualities of American capitalism that value the flexibility of individuals over that of institutions. Americans' admiration of Edison for his sleep, or lack thereof, indexes that American capitalism continues to abide by particular values that stress the contradictory power of individualism and conformity to the spatiotemporal hegemony of the slumbering masses.

The Structure of the Book

This book is divided into three parts: "Sleeping, Past and Present," "Cultures of Sleep," and "The Limits of Sleep." In the first part, which comprises chapters 1 through 3, my concerns are largely historical, and I trace the development of sleep medicine in the United States from its colonial past through its contemporary practice. In the chapters that comprise the second part, I trace the effects of the consolidation of contemporary American models of sleep and its disorders. After presenting a series of cases of individuals and families who confront disorderly sleep, I follow the contemporary politics of sleep into a variety of institutions. These include the workplace, school, and family, as well as institutions that are intimately connected to them: workplace napping facilities, restaurants invested in family values, and corporate work spaces in the United States and around the globe. In the third part, which comprises two chapters and the Conclusion, I focus on the limits of sleep: its place in American legal conceptions of the person, the military–scientific pursuits to eradicate sleep for warfare purposes, and the bioethical acceptance of human variation. These chapters take as their concern the ways that the capacities of sleep have been imagined and the consequences of these scientific, legal, and ethical fantasies for all sleepers, all of which participate in fiction making of various sorts. The Conclusion twists the content of the book into an argument about the ethical basis for the management of sleep and its disorders. I build upon the cases herein to make an argument in favor of *multibiologism*, a politics that accepts the disorderly as human variation rather than medical pathology, which I discuss in detail in the Conclusion. The Conclusion also addresses the therapeutic futures of disordered sleep and their limitations.

What guides these parts are my concerns with doubt in medicine and science, the interrelationship of desire and social formations, and the limits of the human. Although they might seem fundamentally different interests, they are bound together by their basis in the relationship between the individual and the masses; they are processes that both singularize individuals and render them as aggregates, as part of the masses. They instill ideas about the individual and his or her interior powers of self-control as natural and inevitable and simultaneously exteriorize these same powers, rendering them social and patterned. This is the problem of subjectivity, governance, and control, which lies at the heart of the contemporary biopolitical management of life and is at once directed at the discipline of populations as well as focused on the production of individuals as individuals, who conceive of themselves as selves. At the heart of the contemporary biopolitical control of societies is the process of integralism, of individuation and exteriorization, of the individual and his or her relationship to the masses, mediated through the institutions that compose American everyday life: science and medicine, capitalist markets and education, labor and consumption. And at the foundation of this process is doubt: where the individual ends and where the masses begin, what sleep is and what its limits are, and what the capacities of the human are.

Part I is oriented around doubt, first in clinical practice, then by American popular understandings of sleep, and finally in attempts to popularize contemporary sleep science and medicine through media and medical practice. In chapter 1, I begin by describing the contemporary situation of sleep medicine: its rise since the 1970s and the forces behind that rise. In comparison to this expansionism, I describe the sleep clinic where I conducted my fieldwork, which lays the foundations for my analysis throughout the book. Also in chapter 1, I focus on two cases of clinical misdiagnosis (in which patients had been diagnosed as having sleep disorders by other physicians) and how the physicians I studied later rediagnosed these patients. The thread that runs through this chapter is a discussion of the definition of sleep, which is usually based not on what sleep *is* but on what sleep *is not*. Chapter 2 moves my discussion backward in time and traces discourses of sleep from early colonial America through the scientific efforts in the first half of the twentieth century. This may seem a broad swath, but my attention is focused on American idioms of sleep, which highlight concerns with efficiency and productivity. The comprehension of sleep through its metaphors compounds the doubts that motivate clinical practice. Throughout chapter 2, my concerns are with what makes American conceptions of sleep *particular* and how these particularities determine the possibilities of American sleep. Chapter 3 returns to the contemporary sleep clinic, drawing on my fieldwork at the Midwest Sleep Disorders Clinic. I focus on the roles of doubt and certainty in medical practice and on how physicians discern sleep as the cause of disorder in the lives of patients. The foundation for this analysis is a series of cases presented by physicians to other physicians and the discussions that ensued, as well as the media focus on the promotion of sleep as an explanatory device for understanding individual and social disorder. I conclude chapter 3 with a discussion of integralism in medicine, the forces and motives in producing the slumbering masses and its relation to doubt.

Throughout Part II, I am interested in the mechanisms through which individuals and institutions become inseparable, each providing the other's justification for their spatiotemporal ordering. These desirous formations, in which individuals order their lives through the spatiotemporal demands of institutions and in which institutions are ordered through recourses to conceptions of human nature, provide the framework for Part II. I begin by exploring this formational desire and then present a series of cases of individuals and families who identify as specific kinds of disordered sleepers: narcoleptics, insomniacs, sleep apnics, and so on. Each of the kinds of cases appears again later in the book, as individuals and families come into contact with the institutions I subsequently focus on, so my concern in presenting them is in evidencing the conditions under which some individuals and families come to see their sleep as disordered and the possibilities they confront. Rather than offer a conclusion to these many cases, I instead open them up into the following seven brief chapters in Part II. These chapters provide ethnographic descriptions of a variety of institutions that confront sleep and its disorders, and they examine the politics and consequences of these confrontations. Rather than accept the institution of work or school or family as enclosed unto itself, I take seriously that the demarcation of any one of these institutions is always established in relation to other institutions. The impact of one institution on an individual necessarily influences other institutions and other individuals. Instead of focusing on sole institutions or individuals in these chapters, I draw situations together across institutions and individuals—situations that produce particular desires, intimacies, and orders.

In Part III, I take these discussions of doubt and integralism, desire and intimacy, and place them into a dialogue with the limits of human capacity. Chapter 11 traces the history of the use of the sleepwalking defense in the United States and its roots in conceptions of desire and nature, which find their basis in fiction and popular representations as much as in science and the law. I follow this discussion by focusing on the invention of drowsy driving laws early in the twenty-first century, laws that explicitly recognize the effects associated with ongoing sleep deprivation. Conceptualizing the role of culpability in one's actions while sleeping is integral in understanding what makes us human, what the limits and ends of our agency are. How the law interacts with the desires and intimacies of individuals through their culpability in sleep-related actions makes evident how the masses undergird conceptions of individuals and the limits of their humanity. Chapter 12 takes this concern with how sleep makes us human into stranger realms: extreme sports, military research, and adventurous science. At once, the very basis of sleep is being tested in these situations while sleep's potentialities are being isolated. The questions that guide these efforts have led me through a variety of experiments with sleep that bring together those interested in reconceiving the formation of everyday life, the capacities of individuals, and the institutional interactions that make life possible. In the Conclusion, I return to a number of the cases presented throughout the book and put them into a dialogue with the recurrent popular and scientific interest in eradicated sleep. What compels us, despite the seemingly universal experience and presence of sleep on the part of humans and all life on Earth, to want *not* to sleep, to separate ourselves from our nature? I introduce cases of perpetual sleeplessness to challenge this imagined posthuman future of sleep, attempts to break records to stay awake, science fictional representations of experiments to eradicate sleep, and scientific attempts to reduce or dramatically alter our sleep. From the perspective of sleep, tarrying with the simplifications of understanding pathology as variation necessitates a politics of biological inclusion, a rethinking of our human capacities and posthuman conditions of life, a politics of multibiologism.

Sleep is a manifold process, extending its force into our everyday institutions and lives in complex and often unacknowledged ways. Unraveling the complexities that make up our desires for sleep and its intimacies opens up questions about our humanity, the nature of society, and the cultural place of nature in our everyday lives. Moving among the origins of modern American conceptions of sleep in the 1600s, the present, and the futures that are imagined for our sleep exposes the always emerging capacities of everyday life, for both individuals and institutions. Regularity, order, and normalcy become the thresholds for the biology of everyday life as technologies, science, and medicine reformulate what our capacities are and what they might become. What our sleep will become, and what it is, will be measured against our actual and imagined histories and futures; we will always be sleepers, but how we sleep and how sleep influences our everyday lives alter our expectations of normalcy and pathology, of variation and rhythm, and mutate our capacities, changing the very conditions of life and its forms.

I.

SLEEPING, PAST AND PRESENT

1

················

The Rise of American Sleep Medicine
Diagnosing and Misdiagnosing Sleep

DEFINITIONS OF SLEEP WERE PLENTIFUL IN THE NINETEENTH century. For instance, Henry Lyman begins his *Insomnia and Other Disorders of Sleep*, "Natural sleep is that condition of physiological repose in which the molecular movements of the brain are no longer fully and clearly projected upon the field of consciousness."[1] Lyman was adding precision to William Hammond's definition, which he proposed in his *Sleep and Its Derangements:* "The *immediate* cause of sleep is a diminution of the quantity of blood circulating in the vessels of the brain, and . . . the *exciting* cause of periodical and natural sleep is the necessity which exists that the loss of substance which the brain has undergone, during its state of greatest activity, should be restored."[2] Hammond, in his turn, was adding to Robert Macnish's definition of sleep, which appeared in the first modern monograph on sleep, *The Philosophy of Sleep.* Macnish writes, "Sleep is the intermediate state between wakefulness and death: wakefulness being regarded as the active state of all the animal and intellectual functions, and death as that of their total suspension."[3] Compare this with Donald Laird's 1930s discussion of why humans sleep, approximately a century later: "Why people sleep may never be known. And what sleep is may never be satisfactorily stated. Sleep, in these respects, is like life itself, of which it plays so important a part; the biologist cannot tell what life is, but this does not prevent the physician from saving it and prolonging it."[4] One of the characteristics of contemporary conceptions of sleep is that it is inscrutable, that it remains a mystery to even those whose medical practice focuses on the remedy of its disorders.

Contemporary sleep medicine was invented in the 1950s. The beginning date of modern sleep medicine can be easily identified: it occurred when physicians created the category of what was then referred to as "Pickwickian syndrome," named for a character in a Charles Dickens novel, but has been called obstructive sleep apnea since the 1960s. Ten years later, William Dement began to work earnestly on understanding narcolepsy, and his efforts at the Stanford Sleep Disorder Center helped to move sleep medicine further toward disciplinary consolidation. In the late 1970s and early 1980s a host of other sleep-related disorders came to be nosologically defined, and by the end of the 1990s, with the wider availability of pharmaceuticals and prosthetic treatments, sleep medicine was a well-defined subdiscipline of American medicine. The existence of a recognized subdiscipline means many things: institutional and financial support, annual meetings and academic journals, advocacy and ethical review boards, and the need to expand the capabilities of the discipline further. Whereas this attention to sleep within medicine happened in relatively piecemeal fashion from the 1950s through the 1980s, by the beginning of the 1990s, it was intensifying. In this chapter, I examine two formations: the contemporary sleep clinic and its workings, and the rise of sleep medicine in the United States, from the 1930s through the turn of the twenty-first century. In telling this history, I am interested in the ways that normal sleep has been defined implicitly through defining pathological sleep. Sleep has come to be known not by what it is but by what it is not.

Sleep has long been defined by its negative status (it is not wakefulness, nor is it coma or death), and this negative status allows for an interpretive flexibility in the clinical understanding of sleep and its variations. This ambiguity allows negotiation within sleep medicine and science, changing the objective nature of sleep, and also allows for the expansion of sleep's importance in its increasing association with other concerns. Building upon the analysis of sleep's definition, I describe the sleep clinic where I conducted my fieldwork and focus on two cases of clinical misdiagnosis, in which physicians at other clinics, as a result of expansion in the discipline, diagnosed patients as having sleep disorders they did not have, and the later rediagnosis of these patients by the physicians with whom I conducted my fieldwork. I use these vignettes to explore how the rise of sleep medicine, with seemingly easy treatments and straight-forward nosologic categories, has led to a rise of misdiagnosis and to show how sleep physicians are contesting these oversights through recourses to understandings of human biology as variable and pathological. The possibility of misdiagnosis and a disorder's resulting treatment, I argue, is dependent upon the presence of doubt in medicine generally and on the treatment of sleep more specifically; sleep diagnoses have not become more popular solely because of the rise in popularity of sleep medicine. Rather, the fundamentally slippery nature of sleep is what has led to the popularity of disordered sleep as an explanation for aberrant behavioral and physiological phenomena, and it allows for the possibility of widely disparate diagnoses for the same set of symptoms. The certainty that ends these cases, in the form of a new diagnosis, should not be taken as evidence that the patients have been definitively cured. Given the persistent possibility for new symptoms, new difficulties, new treatments, doubt can always return, and the medical drama can continue to unfold.

Two Diagnoses

One of the effects of the growing popularity of sleep medicine as a medical subdiscipline has been the installation of sleep clinics, wings of hospitals, and private clinics that specialize in the treatment of sleep disorders. In the 1970s there were only three such clinics in the United States. By the 1990s, the number had grown to over three hundred. By the turn of the twenty-first century, there were over twelve hundred such clinics. But they vary widely in the services they offer, from strip mall clinics that specialize in obstructive sleep apnea and have one physician with a number of sleep technicians, to hospital-based clinics that bring together physicians from across medical subdisciplines to focus on sleep. The clinic where I conducted most of my fieldwork with physicians was of this latter sort, although over the past decade I've visited sleep clinics at various hospitals and universities and in strip malls. Although they vary widely in the kinds of services they provide and in the breadth of specialties involved in the care of patients, sleep clinics as such share some basic qualities that are based on assumptions about science, medicine, and American sleep.

Sleep clinics are often quiet places, far removed from the center of the hospital's action. But, like the rest of the hospital, they are clinical, institutional spaces, decorated in subtle, pragmatic colors. The MSDC, which requires visitors to take an elevator ride to reach it, is located near the top of the multistory building at the center of the Mississippi County Medical Center. Most people would board the elevators on the first floor, in which case their approach would take them through the admissions area, filled with waiting patients, past a gift shop and snack bar, through unmarked hallways, and eventually to the elevators nestled in the middle of the hospital complex. Visitors not knowing where they are heading are likely to get lost and be redirected by hospital staff. The relative difficulty of finding the MSDC and the other clinics located above the first floor ensures that some random passerby will not stumble into the clinic. Stepping off the elevators, visitors can turn left or right, the former leading to the observation rooms for patients to sleep in, the latter to the main body of the clinic. Walking into the clinic, visitors find themselves at the head of a T-shaped space, first facing an intake desk and small waiting area. The waiting area, like many clinical waiting areas, is supplied with stacks of old magazines, brochures for patients and their family members, and toys for young visitors. Along the main corridor of the clinic are offices for staff, a communal workroom, a conference room where weekly lunches are shared, and rooms filled with accounting and medical files. At the far end, some two hundred feet from the intake desk, is the sole bank of windows allowing sunlight into the public space of the clinic.

The centerpiece of every sleep clinic, like the MSDC, is its assembly of observation rooms for sleeping patients and a control room from which the patients can be monitored while they sleep. Each of the observation rooms is outfitted with a single bed for the sleeper. These beds are hospital beds, designed not for comfort but rather for confined observation: they're narrow, with guardrails on each side, and outfitted with thin blankets and pillows. Around each bed is an array of monitoring devices: sensors that are attached to the sleeper's body to monitor depth of sleep, muscle movement, heart rate, and breathing, infrared video cameras to watch the sleeper through the night, and microphones to capture any noises the sleeper makes. The number of beds in any clinic can range from just a few to several, with large clinics having twelve or more beds. While the sleepers sleep through the night, sleep technicians staff the observation room, with usually one technician for every three or four beds. The technicians float through the observation room and clinic throughout the night. As Sharon, a technician I met during my fieldwork, explained to me during a tour, "The action doesn't really start until 2 A.M. That's when everyone has gotten to sleep and they start snoring or whatever. And then it's a half hour of activity as we monitor people and adjust CPAP machines. The rest of the night is usually pretty quiet, unless there's something severe we're monitoring." The observation technology records the whole night of activity as well as any naps that patients are sometimes asked to take throughout the following day. Before everything was digitized, technicians would have to go through hours of paper-based recordings and highlight portions that indexed symptomatic experiences of the sleepers. Now, everything is recorded through computers, and technicians can scroll through hours of sleep in seconds.

In the morning, physicians examine reports that the technicians have prepared for them, usually to determine the severity of symptoms. How many apnea events did a patient have in what amount of time? How much sleep did the individual complaining of insomnia finally get? Does that patient have REM behavior disorder or night terrors? Throughout the day, patients can be tasked to take naps at two-hour intervals to determine their daytime fatigue. The multiple sleep latency test, developed by William Dement and Mary Carskadon, monitors how long it takes a patient to fall asleep after he or she lies down for a nap.[5] Patients with narcolepsy or severe daytime sleepiness can often fall asleep within a minute; for patients with other conditions, daytime fatigue is less of an issue, and they may be exempted from these tests. On the basis of the night's sleep and whatever occurs during the following day, the technicians and physicians will narrow the diagnosis and prescribe medication or prosthetic treatment. In cases of severe sleep disorders, patients may be asked to spend a second night sleeping in the clinic for the purpose of observing the efficacy of the prescribed treatment. In rare cases, patients might be admitted for longer stays for more intense observation, but most treatments are readily apparent and effective with some adjustment.

Physicians and technicians meet regularly—sometimes daily, sometimes weekly—to discuss cases.[6] First and foremost, they are interested in finding appropriate diagnoses for patients in their care. Secondarily, they are interested in cases that exceed their current understanding of sleep disorders. And finally, they are interested in cases that confirm nosologic categories, disorders that meet the expectations of symptoms and treatment that have been established. With novel patients, the intricacies of the disorder are discussed, with the attending physician presenting the case to the clinic's other physicians and technicians. These presentations are both for collective understanding of the case and to seek insight and advice from those present. The presentation of a case usually includes a brief history of the patient and his or her symptoms. It can often range into more sociological data as well, as the presenting physician might discuss the impacts of the patient's sleep disorder upon her or his social life, including the effects on family, school, and work. Possible solutions are often presented, with a focus on both alleviation of symptoms and remedies relating to social concerns. From there, case discussions can ramble into reviews of recent publications related to the presented disorder, other similar cases that have been seen in the clinic, and broader sociological or philosophical discussions.

Where sleep clinics vary, I have found, is in the latitude they exhibit in this last interest, context, with some clinics rambling across wide intellectual terrain and others focusing quite narrowly on solving the cases of patients. Moreover, the interpretive possibilities vary among clinics, with some narrowly focusing on their specialties and others thinking broadly, not just about sleep disorders, but about physiological and psychological concerns more generally. There are striking resemblances among many sleep disorders, but their differences necessitate treatments that are fundamentally distinct. For example, although they present themselves similarly, primary insomnia and secondary insomnia require different medications, because the treatment for one can fail to resolve the symptoms of the other. The wrong treatment at best is benign and at worst can have significant side effects, compounding the already existing disorder. As a result, some patients can be diagnosed with disorders they seem to have, but only after failed treatments do they return to the clinic to be reassessed for their persistent or emergent symptoms.

Kristy was diagnosed with narcolepsy when she was thirty years old. I met her at age thirty-four through a narcolepsy support group, and we discussed her medical history and its impacts on her life. Her symptoms had been present for about twelve years previous to that and included chronic fatigue and a diminished immune system, leading to frequent colds and bouts of the flu. Since the symptoms occurred at the end of her teenage years, Kristy thought it was "just one of those adolescent things" and that it had to do with aging and general changes in her body. She sought medical help and was first diagnosed with mono. When her symptoms failed to remit, her family doctor thought it might be a severe yeast infection. That was treated, but again her symptoms persisted. The next guess was an adrenal deficiency. This was all happening in the early 1990s. By the time Kristy was eighteen, in the middle 1990s, sleep medicine had reached a more entrenched position in American medicine, and her symptoms then appeared to be associated with narcolepsy. By then, she was in college, and was tired throughout her courses. What seemed to be clear evidence that she was a disordered sleeper was that one day she fell asleep on the strip of grass next to the sidewalk on her way home from class. Kristy would nap daily, ranging anywhere from one to two hours. Her family chided her about it; they thought she should be able to "get over it—get past it." But the symptoms were beyond her control. When she was finally diagnosed with narcolepsy, in 1996, she was treated with two drugs, Provigil and Xyrem; the first is intended to promote wakefulness during the day, and the latter consolidates sleep at night. It seemed that her symptoms were finally being treated, but the Provigil would increase her heart rate to uncomfortable levels. It took a while to get the dosage under control.

As Kristy aged, it seemed that the narcolepsy diagnosis was an appropriate one. Once the right balance of pharmaceuticals was reached, her sleep normalized into a consolidated, nightly stretch. The difficulty came when she got pregnant. Because of uncertainties about the effects of her narcolepsy drugs on the health of a fetus, Kristy went off her medications, and her symptoms returned. This made pregnancy more difficult than it would have been normally, as she struggled to stay awake during the day and slept fitfully through the night. After the baby was born, Kristy had to stay off her Xyrem for fear that she would fail to hear her child cry through the night. Xyrem is a powerful sedative, ensuring four hours of continuous sleep through total sensory seclusion. Her symptoms began to get worse: "profound fatigue, to lack of restful sleep, to getting sick often, to brain fog, to tachycardia, to skin breakouts, to asthma." Maybe, Kristy began to think, narcolepsy still wasn't the right diagnosis.

Chip was referred to the MSDC with REM behavior disorder (RBD), a condition that is known to affect men in their fifties and older and women with narcolepsy. Chip, however, was in his late twenties. Chip's symptoms included a delayed sleep phase—being unable to fall asleep until 1 or 2 A.M.—and sleepwalking. Most troubling was that he would often urinate around the house while sleepwalking and on two occasions was found by his wife in their front yard, urinating. Chip worked in construction, which demanded his awakening at 5 A.M. to be at work by dawn. As a result, he was also chronically sleep deprived. In the words of one of the attending physicians, Chip's "marriage [was] on the ropes." His wife was similarly sleep deprived because she would awaken when Chip did and then follow him around their house to make sure he didn't hurt himself or their four-year-old son. What had led his previous physicians to suspect the correct diagnosis for Chip was RBD was that he also threatened his wife when he was sleepwalking—a behavior widely associated with RBD. What supported this perspective was that Chip was polite and generous with his wife during the day, which made his nocturnal testiness all the more convincingly associated with RBD, because many individuals diagnosed with RBD are genial during the day only to turn violent during sleep. He had previously been put on Ambien for his sleep phase delay and clonazepam for his RBD. But this combination of medications resulted in sleep-related eating behaviors (his most frequent being handfuls of peanut butter scooped from the jar) and sleep-related sexual behaviors, often groping his wife in his sleep, which understandably unsettled and angered her.[7] Because his symptoms continued, increasingly disrupting his family, Chip continued to seek medical help for his sleep disorder. RBD seemed to be an appropriate diagnosis, given his symptoms, but upon closer scrutiny his disordered sleep was both more complex and more troubled than such a straightforward diagnosis allowed for. Understanding what was affecting both Chip and Kristy necessitates attending to sleep as it is currently conceived, as well as its pathological forms. What might be unsettling their nightly sleep?

Defining Sleep and Its Disorders

"I define sleep in terms of only two essential features," writes William Dement in his book *The Promise of Sleep*. "The first, and by far the most important, is that sleep erects a perceptual wall between the conscious mind and the outside world. . . . The second defining feature of normal sleep is that it is immediately reversible. Even when someone is deeply asleep, intense and persistent stimulation will always awaken the sleeper. If not, the person is not asleep, but unconscious or dead." Dement adds, "It occurs naturally . . . and it occurs periodically."[8] Trained by Nathaniel Kleitman at the University of Chicago in the 1950s, Dement was of the first generation of sleep researchers and physicians who accepted that sleep was an active period, not one of total bodily cessation. In his magnum opus, *Sleep and Wakefulness,* Kleitman had previously written against the idea that sleep was "a periodic temporary cessation, or interruption, of the waking state." Instead, Kleitman argued, "Sleep is in reality a complement to the waking state, the two constituting alternate phases of a cycle. . . . Without wakefulness sleep cannot be said to exist."[9] It may appear commonsensical now to think of sleeping as being active, but Kleitman had initially posited sleep as active thirteen years before the discovery of rapid eye movement (REM) sleep. The state of the art was still Sigmund Freud's *Interpretation of Dreams,* in which Freud had posited that dreaming provides a means to ensure sleep (it consumes the sensory apparatus that might otherwise be distracted by nighttime stimuli), and dreaming seems to be sleep's primary, if not sole, activity.[10] The discovery of REM sleep radically upset the presumption that sleep is something that occurs only in the mind.

In 1953, while a graduate student in physiology at the University of Chicago, Eugene Aserinsky was the first to record evidence of what would come to be called REM sleep, also known as "paradoxical sleep" (so coined by Michel Jouvet).[11] That, at least, is the simple, uncontroversial version of the story. Aserinsky had enrolled as a graduate student having earned only a high school diploma; in the post–World War II years, he benefited from the GI Bill and, despite his failure to earn a bachelor's degree, was able to convince the University of Chicago to admit him as a Ph.D. student. He began working with Kleitman, although Aserinsky had no special interest in sleep. Rather, as all narratives of the discovery of REM point to, Aserinsky was primarily concerned with completing his degree and finding gainful employment; it hardly seemed to matter what he studied as long as he was financially supported and marketable upon graduation.

Kleitman dispatched Aserinsky on a project to study the eye movements of children while falling asleep, curious about the point at which children stop blinking and begin sleeping. Kleitman's assumption was that blinking would indicate wakefulness, and when the child had fallen asleep all eye movement would cease. Aserinsky recruited his preadolescent son as a participant and wired him to an early polygraph machine to record the motion of his eyes while approaching sleep. He noticed that after sleep onset, eye movements continued, first as a gentle rolling of the eyes—something that had been documented as far back as the 1820s by Robert Macnish and is now referred to as non-REM sleep (NREM)[12]— and then as a "jerky," rapid movement. In 1953, Aserinsky coauthored a paper with Kleitman detailing the discovery of REM sleep,[13] but years passed without much notice paid to the potential of this finding, namely, that our bodies continue to be motive in perceptible ways even while our perceptions of the world are limited.

Throughout the 1950s, William Dement assisted Aserinksy, a fellow graduate student, in his research. Dement continued the study of REM sleep when Aserinsky left the University of Chicago upon graduation to explore the possibilities of altering the movements of salmon with electrical current at the University of Washington.[14] Although Aserinsky is properly attributed with the discovery of REM sleep, Dement pursued the ramifications of REM sleep doggedly: inspired by Freud's work on dreaming, Dement saw in REM sleep a means to study the dreaming mind, because it appeared that REM sleep was proof that sensory processing occurs during sleep, although it focuses on dream content instead of what is occurring in the waking world. Dement conducted a variety of experiments in an attempt to prove the linkage between REM sleep and dreaming. By attaching research subjects to polygraph machines that detected the minute alterations in electrical activity generated by muscle movements, he was able to detect precisely when a sleeper had entered REM sleep. The researchers would gently awaken the sleeper and ask about any remembered dream content. With some 75 percent of sleepers reporting that they had been dreaming when their eyes were recorded as being in movement, Dement became convinced that the linkage between REM sleep and dreaming is absolute; it was only years later that dreaming was proven to occur in many, if not all, of the sleep phases.[15] Moreover, Allan Rechtschaffen, who had assumed Kleitman's informal position of primary sleep researcher at the University of Chicago upon the latter's retirement in the early 1960s, demonstrated through a number of experiments that the content of dreams is almost wholly internally produced, rather than somatically or environmentally influenced.[16] Rechtschaffen and his students innovated a way to prop eyelids half open and still ensure that research subjects fell asleep. After a test subject had entered REM sleep, Rechtschaffen would enter the sleeper's bedroom and illuminate objects held before the sleeper's half-open eyes. He would then awaken the dreamer to see if the illuminated object had been translated into the imagery of the dream. Almost invariably, the objects had been ignored, and when the dreamer had processed them, they would assume only secondary importance to the dream. The result of this was that sleep science entrenched the idea that sleepers are enclosed unto themselves, ensconced behind Dement's "perceptual wall."

This might seem a slight transformation in understanding sleep, but its implications were profound. The understanding that the sleeper is an essentially closed system provided the emergent science and medicine of sleep with two foundational components: first, this tenet disconnected sleeping from the sensory world, rendering sleep a separate terrain for action; second, it made sleep disorders primarily individual, extricating them from their social context. This is the twin basis of contemporary sleep medicine: sleep exists outside external influences, and its disorders are individual. If these claims are readily apparent, it is because these tenets of sleep medicine have entered dominant understandings of sleep. These claims stand in opposition to how sleep was considered and acted upon previously. Taken together, these claims provided the means to target treatments at an individual's physiologies—not at psychological stresses or environmental conditions, but at biology. As a result, pharmacological and technical solutions could be produced to alleviate the biological disorders of individuals. This required that the etiology of a disorder be isolated so that a therapy could be developed to address the identified cause.

These two assumptions were added to another, namely, that consolidated sleep is the human norm. This will be discussed in greater depth in chapter 2. For now, it is important to note that over the course of the twentieth century the idea that sleep occurs in one consolidated eight-hour period, preferably nocturnal, came to be accepted as scientific fact. In order to achieve this consolidation, a new understanding of fatigue needed to become central to the conception of sleep. That is, the idea that an individual will experience being tired regularly was normalized; exhaustion at the end of the day without a nap is quite different from exhaustion that is experienced despite a daily rest. Dement writes, "[A] small amount of sleep debt is good, indeed is necessary, for sleeping efficiently."[17] Fatigue became "sleep debt," and "efficient" sleep became of paramount clinical concern.

These economic understandings of sleep and its disorders should come as no surprise. The consolidated model of sleep is predicated upon the solidification of other institutional times in American society, foremost among them work time. Consolidated sleep provided a clinical model to train aberrant sleep toward. This might seem to be necessary only for insomnia, but all sleep disorders impact consolidated sleep and often confound its "efficiency" by rendering it neither deep nor sound enough. Even for normal sleepers, the circadian cue for sleep onset advances throughout the week as a result of a dyssynchrony between physiological prompts and institutional temporal ordering. As a result, the impulse to sleep occurs later each day, as does the cue to wake up. Fatigue builds throughout the day and over the course of the week, leading to steady feelings of sleepiness and consolidated sleep through the night. "Debt" leads to "efficient" and consolidated sleep and makes sleepiness vital for wakefulness.

With these foundational assumptions in place, Dement and his colleagues were able to turn toward popularizing the medicine of sleep. In the early 1970s, the emergent understanding of narcolepsy as a physiological rather than psychological disorder proved promising, and Stanford University hired Dement in 1970 to head their newly formed Sleep Disorders Center. Coextensively, Pickwickian syndrome (so named in 1956) was becoming more fully understood, which resulted in the nosologic definition of sleep apnea by 1965 and the eventual production of prosthetic technologies to aid in the alleviation of apnea symptoms.[18] In 1972, when Dement taught his first continuing medical education course on sleep disorders, the field of sleep medicine had officially come into being: "Because a discipline does not exist unless it can be taught, I have suggested that the date this course began might be recognized as the official birthday of the clinical, scientific discipline of sleep disorders medicine."[19] It was believed at the time that narcolepsy would prove to be the primary focus of sleep medicine, but over the ensuing decade, sleep apnea would be recognized as the financial foundation for the practice of sleep

the collapsing of the airway during sleep: the sleeper makes increasingly loud, choking noises during sleep as his or her airway constricts, until such a point that he or she spontaneously wakes up from suffocation. Upon the person's falling asleep again, the cycle repeats itself, leading to chronic sleep deprivation. As a result of the repeated lack of oxygen and loss of sleep, secondary symptoms of OSA include loss of memory, excessive daytime sleepiness, weight gain, and shortness of temper. OSA was first diagnosed as Pickwickian syndrome in the 1950s, but it was in the 1960s that the disorder came to be more identifiable as a disorder in itself and not a by-product of obesity. There is a cyclical relationship between OSA and obesity, since the latter predisposes individuals to develop the former. Because of sleep loss, many people offset their lack of energy with increased caloric consumption, leading to weight gain. This, in turn, exacerbates the symptoms of OSA. The cycle can be interrupted through treatment, of which there is one reliable method and one experimental. In 1981, Colin Sullivan, an Australian inventor, created a continuous positive airway pressure (CPAP) machine, which, when attached to a sleeper by a facemask held in place by an adjustable band, provides air pressure to inflate the airway. This allows the sleeper to breathe normally, albeit with prosthetic help. This technology was revised in the 1990s with the invention of the bilevel positive airway pressure (BiPAP) machine. The BiPAP machine simulates breathing and alternates between two levels of airway pressure. CPAP and BiPAP machines are sometimes difficult for individuals to tolerate, and compliance with their prescription has become a concern for many medical professionals. In order to be effective, the machines need to be used nightly, and noncompliance with this rhythm can lead to symptomatic relapse. For those who cannot tolerate CPAP and BiPAP treatment, an experimental surgery to implant pillar supports exists. Pillar implants support the airway, providing a means for less obstructed breathing. Unfortunately, for most individuals, pillar implants fail to reduce symptoms associated with OSA significantly.

Although the diagnosis of OSA is relatively new, its symptoms are quite old and have been known to physicians for some time. In comparison, insomnia is one of the oldest sleep complaints. What made insomnia new to the twentieth century was finer understandings of its many forms as well as two waves of chemical treatments for the disorder. Insomnia refers to at least four different, clinically recognized disorders: primary and secondary insomnia, sleep maintenance insomnia, and fragmented sleep. What these four disorders share is lack of willful sleep—being able to sleep when one desires it. However, they are significantly different and often require different kinds of medical treatment. Fragmented sleep takes the form of multiple awakenings throughout the night. The awakenings can be unpredictable, and their random quality is what sets them apart from other forms of insomnia. Primary insomnia refers to difficulty with sleep onset. Sleep maintenance insomnia often takes the form of awakening in the middle of the night and being unable to return to sleep, whereas secondary insomnia is usually the result of another physiological problem. The causes for each of these disorders can vary, including such phenomena as stress, anxiety and depression, caffeine and alcohol consumption, and underlying physiological conditions. The primary symptom of insomnia, sleeplessness, is the only rationale for treating these distinct conditions similarly. Because the causes of insomnia are both diverse and common, most adults have periods of sleeplessness that might qualify as insomnia. When the symptoms fail to resolve themselves, medical treatment is often necessary and frequently involves behavioral modification—what is often referred to as "sleep hygiene"—or pharmaceutical interventions or both. The recent attention to insomnia has been due to the Z-drugs, which seem to offer one solution for four distinct conditions. The Z-drugs may act to obscure the differences between the forms of insomnia, producing side effects to the drugs, particularly parasomnia behaviors.

Two sleep problems are accepted as disorders because of their primarily social effects, delayed sleep phase syndrome (DSPS) and advanced sleep phase syndrome (ASPS). The former is often perceived as more problematic, although both are disruptive to those with the symptoms. DSPS is characterized by an inability to fall asleep at an appropriate or desirable time of night, often perceived as insomnia, but the symptoms fail to be alleviated when treated with standard sleep aids. As one might expect, those with ASPS symptoms go to sleep at earlier-than-normal hours, to awake well before the school or work day. They are the proverbial early birds, able to take advantage of the day, but find themselves wasted by the day's end, leaving them unable to attend to the social demands of evening and night. In both cases, what is medically mandated is the shifting of patients' circadian clocks, a complicated process that is rarely successful. This process employs a variable chemistry of melatonin supplements (which have been shown to affect the circadian rhythm of individuals), lightboxes, sleep restrictions (no napping, waking at set times), and dietary restrictions (especially pertaining to sugar and caffeine). In many cases, despite the necessary coordination of these factors in an attempt to regulate an individual's sleeping patterns, if an individual stops the use of the prescribed daily regime, the unwanted circadian rhythm assumes primacy once more.

The parasomnias (translated as "alongside sleep") are internally diverse and include such phenomena as sleepwalking, sleep talking, sleep-related eating disorder (SRED), and sexsomnia. Sleepwalking has a long history, and, as I discuss in Part III, early depictions of sleepwalking are central to American understandings of culpability and agency. Sleepwalking has long been understood as having a genetic basis, and dominant scientific models of sleepwalking have embraced this hereditary basis for the behavior. More generally, the parasomnias have long been popularly understood as arising because of some internal difficulty on the part of the sleeper, whether it's nightmares, poor digestion, or, latterly, psychological trauma. Increasingly, they have come to be understood as the result of an only partly sleeping nervous system. Whereas earlier conceptions accepted that sleep is a "global" phenomenon, with sleep affecting the whole nervous system equally, more recently sleep has begun to be understood as "local."[20] According to the local model, the parasomnias occur as a result of some parts of the brain being asleep while others are awake. This can occur for any number of reasons, including physiological and chemical causes. As a result, individuals can act out in dreams or express behaviors that are uncharacteristic of them in their waking lives. The parasomnias have been central in finer understandings of the nervous system and its relation to sleep, particularly since the introduction of the Z-drugs. Because the Z-drugs act upon the nervous system differently from earlier benzodiazepines, they can sometimes promote parasomnias: they put the body to sleep, but parts of the nervous system may still be active. Insomniacs have been treated with Z-drugs only to find that they suddenly develop sleep-related eating or sexsomnia behaviors, as in Chip's case. Newer benzodiazepines are also used to treat parasomnias; in addition to their sedative effects, they can serve as anticonvulsants and muscle relaxants, thereby reducing motor control and inhibiting parasomnia behaviors. If the parasomnias have helped to reconceptualize the very foundations of what sleep and wakefulness are, narcolepsy provides a model for sleep and contemporary American everyday life.

Narcolepsy (literally, "sleep seizure") was first identified as a discrete sleep disorder in the 1880s. A sudden onset of sleepiness is the primary symptom of narcolepsy, and this burst of sleepiness is often associated with intense emotional feelings. As such, narcoleptics fall asleep throughout the day. The irony of narcolepsy is that it often impairs desired sleep as well, fragmenting it so that individuals rarely sleep through the night. The most common additional symptoms associated with narcolepsy are cataplexy and hypnogogic hallucinations, the former being the sudden loss of muscle tone associated with sleep onset, the latter referring to waking sensory confusion. Some have suspected that narcolepsy is caused by an infection that triggers a reconfiguration of the thalamus and hypothalamus. Others have suspected that the disorder is genetic. Still others conjecture that it might be a balance between the two, a genetic predisposition that reacts to an infection in a way to produce the symptoms of narcolepsy.

The disorder is treated with a variety of pharmaceutical agents, each of which targets one of the disorder's symptoms. Provigil (modafinil) is used as a stimulant throughout the day to relieve sleepiness, sometimes replaced by Concerta or Ritalin, both of which are stimulants that act on the nervous system in a fashion quite differently from Provigil, often resulting in addiction. At night, narcoleptics often take Xyrem (sodium oxybate), a controlled substance sometimes known as gamma hydroxybutyric acid (GHB), notoriously used as a "date rape" drug. Because of that status, Xyrem is distributed through Orphan Medical, a pharmacy that tracks prescription and use of controlled substances. For narcoleptics, Xyrem helps to consolidate sleep into four-hour periods; it also, unexpectedly, reduces the incidence of cataplexy events for many narcoleptics. Despite narcolepsy's relative scarcity in the human population, this need for medical intervention for both sleep and wakefulness makes narcoleptics a model of the modern American sleeper.

Rapid eye movement behavior disorder was recognized as a formal nosologic category in 1986. As a disorder, RBD is one of the more severe, if not the most severe, of sleep disorders. By day, most individuals who are diagnosed with RBD are calm, gentle, and patient. By night, RBD causes individuals to punch, kick, and jump out of bed regularly, and when asked about the content of their dreams, individuals are almost always able to relate a dream in which they were being chased or in conflict with a stranger or a dangerous animal. "I saw two deer go by in the haymow," begins Mel Abel's recounting of a paradigmatic RBD dream,

> and I told my grandpa, "Did you see those deer?" He said, "No, where did they go?"' I said, "They must have gone to the other end of the barn." He says, "I'll go down and roust them out," and I said, "I'll wait here with the pitchfork and maybe I can get the doe." All of a sudden, here that doe came, and I bashed her as hard as I could across the neck and down she went and laid there and blatted. "I know how to fix that up; just get you by the chin and head and snap your neck." I reached over—and I got [my wife] by the chin, and I just put my hand on top of her head, and she let a holler out of her and jumped out of bed.[21]

Abel also described how inanimate objects would often bear the brunt of his sleeping fury, as he destroyed lamps, dressers, bedside tables, and bedroom knickknacks. RBD is incredibly rare, with generous estimates at less than 1 percent of the human population. For years after its initial reporting, it was colloquially referred to as the "Minnesota disease" by sleep researchers, because no one outside the Hennepin County Medical Center (HCMC) or the Mayo Clinic, both located in Minnesota, had diagnosed a case. Despite the low number of RBD cases, the disorder's presentation is disproportionately male, with informal estimates of its distribution being four to one in favor of men. RBD is accepted as a parasomnia, but because of its severity, its effect on only particular populations, and its treatment with a relatively low dosage of clonazepam (Klonopin, a benzodiazepine), it is seen as a distinct nosologic category. Since the early 1980s, the physicians and psychiatrists at HCMC have tracked RBD, but its isolation as a discrete form was complete only when a particular treatment was identified, thereby isolating both its particular symptoms and a necessary pharmaceutical means of alleviation.

Restless legs syndrome has recently come to the attention of most Americans through advertising campaigns related to Requip (ropinirole), a drug originally used for Parkinson's disease but found to be effective in minimizing the effects of RLS as well. RLS was first isolated as a discrete disorder by Karl Ekbom in 1945, but not until the 1980s did it garner significant attention by clinicians and researchers. The symptoms of RLS come primarily in the form of somatic sensations in the legs, although arms can be affected as well. To relieve these feelings, individuals move their limbs. When a bout of RLS symptoms occurs during the day, individuals might be able to walk around, swing their arms, or shake their legs under a table. At night, while lying in bed, relief cannot be had so easily. Instead, individuals shift their legs back and forth, rub them together, and are generally uncomfortable as they struggle to fall asleep. Currently, most sleep researchers accept that the cause of RLS is either a dopamine or an iron deficiency, and evidence suggests that it may also be largely a hereditary illness. On the basis of its perceived causes, dopamine agonists, such as Requip, can be used to minimize symptoms. When RLS is the symptom of a blood disorder such as anemia, it can often be treated with iron supplements. Because the symptoms of RLS are especially apparent when an individual is at rest or relaxed, this condition has come to be recognized as a disorder of sleep, although its symptoms are present throughout the day and might otherwise be treated as a complaint not specifically associated with sleep. Its status as a sleep disorder is due in no small part to the successful marketing of Requip, which focuses its advertisements primarily on disrupted sleep caused by RLS—for the sleeper and his or her bed partner.

Fatal familial insomnia (FFI) is a severe form of chronic insomnia that is assumed to be largely hereditary. By all accounts, FFI is an incredibly rare prion disease and is genetically transmitted through about thirty family lines, located primarily in Europe and the United States. As such, it has rarely been clinically observed in detail, and no cure for it has been identified. Most often, patients are admitted for care only once the disease has progressed to such a point that they no longer sleep, and their deterioration quickly accelerates thereafter. According to the cases clinically observed, the onset of FFI usually occurs between age thirty-six and sixty-two. From the time of diagnosis, individuals can live anywhere from eight months to six years. Because the disease initially presents itself as insomnia, it can often be ignored for the serious problem it is until it reaches more extreme stages of development. FFI leads to a severe decrease in individuals' sleeping times through the night. As a result, individuals lapse into sleeplike fugue states throughout the day, which can include parasomnia behaviors, such as sleepwalking and dream enactment. Throughout this intensification of symptoms, FFI also makes individuals resistant to traditional sleep aids. Pasquale Montagna describes the advanced stages of FFI in the following way:

> Later stages are characterised by ever-increasing oneiric stupors (from which patients are awakened with difficulty) persistent drowsiness and myoclonus, inability to stand and walk, increasing dysarthria and dysphagia, and loss of sphincter control. . . . Some patients die suddenly in full consciousness and others lapse into a vegetative state, with death occurring because of respiratory or systemic infection.[22]

Like RBD and its relation to the parasomnias, FFI, in its similarity to insomnia, can appear to be a manifestation of what is commonly assumed to be a relatively benign, if uncomfortable, disorder. However, as Montagna notes, as FFI develops it becomes life threatening and radically transforms the individual into a waking sleeper.

Kleine-Levin syndrome is in many respects the opposite of FFI. Sometimes referred to colloquially as "Sleeping Beauty syndrome," KLS's primary symptom is protracted periods of intense sleepiness. Individuals will go for days, if not weeks and months, of sleeping ten hours or more every day, sometimes in excess of twenty hours. During these periods of intense sleepiness, individuals will awaken only to eat and go to the bathroom; hyperphagia (excessive eating) has been associated with KLS. When awake, individuals who exhibit KLS symptoms are often in fugue or confused states, and many also evidence hypersexual or depressive behaviors. KLS largely affects young men, often in their teens, terminating when an individual reaches his early to midtwenties. It can also affect adults, and when it does it fails to resolve itself so neatly. Some studies conducted with the very small population of individuals who exhibit KLS symptoms have shown that symptoms persist for years and can vary in their intensity throughout the course of the disorder. At first glance, KLS appears to be a problem of wakefulness. But when individuals with KLS are treated with drugs to promote their alertness, the best result is often only a shortening of the sleep period; the hyperphagia and the confused waking persist. Because of its cluster of symptoms, KLS can take years to identify as the disorder underlying the symptoms; instead, individuals can be diagnosed with bipolar disorder, thyroid conditions, and narcolepsy. However, even when KLS is finally diagnosed, there is little that can be done, other than attempting to assuage the social tensions that have arisen as a result of the presence of the KLS symptoms.

Shift work sleep disorder may be the only distinctly "modern" sleep disorder, relying as it does upon the frequent alternation of work time from morning to evening shifts. Police, firefighters, nurses and other medical professionals, pilots, and other workers who have jobs that require them to change their schedule dramatically from one week or month to the next frequently develop insomnia, inopportune sleepiness, and often more severe mood-related side effects, such as anxiety, stress, and depression. Workers find that after they have worked a period of night shifts, often working late into the night or until dawn, they have difficulty sleeping during the day. When they start working days again, they have difficulty sleeping at night, having adjusted somewhat to the new requirements of sleeping through the day. When treated as insomniacs, many individuals with symptoms related to SWSD find no relief in the use of sleep aids. Instead, individuals who have SWSD symptoms are often given wakefulness-promoting drugs, such as Provigil or Nuvigil (armodafinil) to help them get through their workdays. The logic behind this is precisely that which underlies Dement's understanding of fatigue as fundamental in the production of consolidated sleep: through consolidating wakefulness, sleep will occur at appropriate times.

This wide variety of disorders that sleep medicine can treat provides the subdiscipline with profound powers. When one considers how each of these disorders is a disorder of the everyday, the relief of even the slightest sleep complaint makes significant differences in the lives of individuals and their families. Moreover, with life-threatening conditions such as FFI and RBD, sleep medicine has the potential to save lives. From the outside, the difference between the symptoms of sleepwalking and RBD, or between insomnia and FFI, might seem slight. But for individuals and families experiencing these symptoms, the differences are profound. For the general practitioner or nonspecialist, again, the distinctions between symptoms might be imperceptible, perhaps because the variations in symptoms are slight, and the ability to discern their differences is central to the power of sleep medicine practitioners. Sleep and its disorders are inherently slippery because of their basis in the definitional negation of sleep. Even for specialists in sleep medicine, discerning one sleep disorder from another is not straightforward, and misdiagnoses occur, in no small part because of the negativity of sleep, the understanding of it as not waking, not coma, not death. In the next section, I return to the two individuals discussed above, Kristy with her narcolepsy and Chip with his RBD. Their respective misdiagnoses have as much to do with the complexity of their symptoms as with the negativity of sleep.

Two Diagnoses, Redux

The straightforward way to interpret both Kristy's and Chip's misdiagnoses is that they had seen physicians who were incompetent. Or, alternatively, they bore the brunt of popularizing sleep medicine; a sleep disorder diagnosis was simply ready at hand because of the increased attention sleep had been receiving. This especially seems to have been the case with Chip: the diagnosis of RBD that he received was clearly based on an oversight on the part of his attending physicians, who ignored the evidence that RBD affects only men in their fifties and older. Even when Chip sought further medical help at the MSDC, his diagnosis was far from simple. Given his complex social and physiological symptoms, isolating the cause from the effects was clinically difficult. In the discussion of his case, one of the physicians at MSDC suggested that Chip be allowed to "free run," that is, be allowed to sleep as he wanted to for a set length of time, without the burden of work responsibilities or family life. This would allow his symptoms to be isolated, indicating if delayed sleep phase was the primary problem, or insomnia, or something else entirely. Allowing Chip to free run was impossible, however, because his family depended on his steady income, and his taking two or more weeks off work would have resulted in either his dismissal or unfavorable work shifts when he returned. The best that could be done for Chip was to attempt to alleviate his symptoms. This end was eventually met. What it required was working through the possibilities that Chip's symptoms evidenced. Physicians prescribed a series of medications in order to find the right combination of drugs to address his physiological needs and social obligations. RBD may not have been the most likely cause of his symptoms, but given the presentation of his disorder, RBD was a possible explanation for his nocturnal behaviors. What was actually affecting Chip was never discovered, but a remedy was found nonetheless.

Kristy's case is much more complicated. Although her symptoms appeared to be those of narcolepsy, they were actually produced by corn, soy, and peanut allergies. As Kristy explains:

> My food allergies do not result in immediate reactions as one would stereotype. They are instead delayed reactions. I may feel the reaction up to a week later. And, it can affect any and all body systems. It's different for each person. In my case, if my food even touches the allergens, I normally react in approximately three days with symptoms of fatigue, aches, brain fog, headaches, et cetera. This reaction lasts for a few days.

Unknowingly exposing herself to each of her allergens through her diet and use of cosmetic products, Kristy constantly experienced symptoms that appeared to be narcolepsy. Kristy had been tested for allergies previously, but since she had no immediate reaction to the allergens, she dismissed their effects on her system. After being retested, her physician recommended that she avoid all contact, even cursory, with allergens. Kristy adds, "Within a week of taking these items out of my diet as well as eliminating products such as skin care products and deodorants containing these ingredients, I felt like a new person! I would not have believed it myself if I had not experienced it." Is clinical incompetence enough to explain the misdiagnosis of Kristy as a narcoleptic? Or is the rising popularity of sleep disorders a means to understand symptoms?

The success of the underlying assumptions of sleep medicine—that sleep is a discrete terrain of medical action and that sleep is inherently individual and relates to particular physiological disorders—provides the means to understand why Kristy labored so long under a misdiagnosis. If disordered sleep is a problem, it provides a delimited space in which to identify and treat a pathological concern. From all of the possible problems that might develop in a human body, sleep radically limits the set. However, within the set of sleep disorders, minor variations mark significant differences in the somatic experience of individuals, their treatments, and the eventual cessation of symptoms. The difference in diagnoses among sleep disorders may also serve as a conceptual trap: discerning the differences among diagnoses and treatments can prove challenging, as evidenced in Chip's case. Once patients are diagnosed with a sleep disorder, physicians can find it difficult to recognize that the set of symptoms is anything other than a sleep disorder; if the patient doesn't respond to treatment, then the assumption is that the disorder must not be the one he or she is being treated for. This doubt is integral to sleep medicine and to allopathic medicine more generally. It serves simultaneously as centripetal and centrifugal forces, offering both an explanation and a lack of resolution. Sleep disorders can be properly diagnosed and solutions posed, as the history of sleep nosologies evidences. In the case of any solution, however, the treatment has only begun: the ongoing use of pharmaceuticals and prosthetics is often necessary to resolve symptoms. Therapy produces an ever-receding horizon of normalcy, which treatment is accepted as helping to achieve but is necessarily elusive. For both Chip and Kristy, their final diagnoses entailed new modes of life, in the chemical and social exposures their bodies were subject to as well as in the organization of their everyday lives.

Again, we know sleep through what it is not. We know it as a discrete physiological function because of what it is defined against: death, coma, and wakefulness. And we know normal sleep through its pathologies, all of those nuisances and aberrations that interrupt a night's sleep or make it impossible altogether. But to recognize pathologies, we need a stable foundation of what normal sleep looks like, as in the consolidated eight hours of sleep most Americans desire every night, sound and uninterrupted. At once sleep is specific (defined by what it is not) and expansive, claiming a toehold in disorders that might otherwise be assessed as primarily neurological, pulmonary, or psychiatric. This might seem only appropriate; sleep is, after all, one-third of our daily lives, and our ability or inability to sleep has profound effects. But this medical and scientific attention to sleep as central to our everyday lives and desires is relatively emergent and is due in part to the recognition that even if sleep cannot be properly defined, it can be known and manipulated. Medicine's inclusion of nosologic categories now ensures that physicians, patients, and their bed partners continue to seek out medical interventions for disorders that are understood as arising from abnormal sleep. American conceptions of sleep have changed dramatically over the past thirty years, and these transformations have been indebted to long-standing cultural expectations of what sleep is and is not, what it tells us about individuals and populations, and how it might be controlled.

2

. . . .

The Protestant Origins of American Sleep

The early bird catches the worm.

—William Camden

Early to bed and early to rise, makes a man healthy, wealthy, and wise.

—Benjamin Franklin

A MERICAN SLEEP HAS ALWAYS BEEN TIED TO CONCEPTIONS of death and the virtuous life. Take, for example, the thought of Cotton Mather, with whom so many American attitudes are founded.[1] In *The Serviceable Man,* Mather's paean to the importance of earthly works inspired by God, he writes:

> Who are we to do these things?
>
> Now, in general, here is work for us all. We ought every one of us to serve our generation, before we fall asleep, or it will be but an uncomfortable sleep that we shall fall into. We are, *in Publico Discrimine,* and that man is a wen or a scab, rather than a member of this body politic, who shall decline the service of his countrey.[2]

Mather's use of *sleep* is metaphorical, associating it with death. Our daily and lifetime activities should be in the service of our generation and are vital to our continued belonging to the body politic. At the end of the day, at the end of our lives, we will be judged for those works we have wrought in support of our community. This is similar to his usage in his other sermons in which he equates sleep with the avoidance of divine service and a lack of consciousness of one's earthly obligations.[3]

In this chapter, I trace American ideas about sleep from early colonial discourses among the Puritans, through Benjamin Franklin and his revolutionary contemporaries, to the rise of industrial-era physicians, who took sleep and its ordering as one of their primary concerns. I end with a discussion of early twentieth-century sleep scientists and the turn toward the pathologization of certain forms of sleep. This discussion prefigures later ones about the ordering of the American day and moral discourses regarding the management of sleep; it also provides a deeper basis for understanding the contemporary practice of sleep medicine in the United States and the doubts that motivate clinical practice. In providing this foundation, this chapter focuses in its conclusion on the continuation of particularly American sleep concerns in the twentieth century, namely, Nathaniel Kleitman's research focusing on circadian rhythms and the invention of "fatigue" as integral to modern sleep in the work of William Dement. These twin expectations lead to alternative management of sleep (siestas, midday naps) as cultural pathologies contrary to the expectations of American sleep medicine. In so doing, they help to establish rigid ideas about everyday American life and the need to align individual biological experiences with it.

From the roots of the American nation in its colonial period through the rise of allopathic medicine in the 1800s, sleep has been associated with inefficiency, power, and individual habits. Taken together, these cultural conceptions of sleep posit life as primarily concerned with productivity, with labor. To be not laboring, to be incapable of production, is to fail to realize one's earthly duties—a sentiment formulated by Cotton Mather in colonial New England. And to increase one's efficiency is to increase one's capacity for life. In this chapter, I trace these interwoven understandings of sleep as they conceptualize American everyday life as primarily concerned with labor and productivity. Sleep may be a period of inactivity, of nothingness, but it can be maximized to increase one's capacity for labor.

Isolating what is particularly American about sleep and sleep medicine is a difficult thing. The nineteenth-century transatlantic exchange of ideas and expertise was facilitated by the rise of the printing press, and it allowed for the formation of allopathic medicine as a discipline. Few physicians published monographs, dissertations, and articles on sleep in the nineteenth century, yet by the 1880s a robust body of literature existed in English, German, and French, and texts were often translated from one language to another. With all of this transatlantic exchange of ideas and people, how can the American particularity of sleep medicine be isolated? I move from nineteenth-century thinkers to their successors in the twentieth century, Donald Laird and Nathaniel Kleitman. By the time early twentieth-century sleep scientists were developing their frameworks, they were very clearly drawing on assumptions of earlier sleep science and medicine, as is evident in their understandings of sleep and their use of metaphors and terminology. Early twentieth-century sleep scientists, and eventually physicians, built upon a shared, transatlantic genealogy of thought to make sleep American by intensifying interests in its efficiency and management. Building upon this consolidation of scientific thought, physicians like William Dement turned these cultural assumptions into the basis for medical practice. This was no teleological history in which ideas unfolded in a linear and logical progression to arrive at an endpoint of rational scientific thought. Instead, vague structures of feeling solidified into a hegemonic conception of sleep through the widespread, popular acceptance of sleep as being primarily about life and productivity.

In this chapter, I follow a fairly straightforward path, from Cotton Mather in the late 1600s through William Dement in the late 1900s. The texts I focus on here may not be the most important or influential texts of their age, but they are representative in that they all evidence the particularity of American thought regarding sleep. Bringing this set of texts together—texts that span hundreds of years—makes American interests in sleep and its functions quite evident. And although these cultural conceptions of sleep remain fairly constant, despite their increasing attenuation as understandings of both human physiology and thermodynamics change, scientific and medical conceptions of sleep and its functions change substantially.

Roughly, prior to the twentieth century there were two schools of thought: One based sleep on attentiveness and the senses, and when the senses have absorbed enough stimulation for the day, sleep results as a means to clear the proverbial slate. In contrast, the other ascribed the onset of sleep to changes in circulation, finding support in the apparent linkage between the increased need for blood flow to the digestive system during and after meals and the impulse to sleep. This suggests that blood moves away from the brain to the gut, thereby inducing sleep. In both cases, an external cause is associated with the desire for sleep, whether it be sensory in the broad sense or sensory as it specifically relates to nutrition. Both imagined homeostasis that is interrupted and brought back into balance through sleep's mediating force. Not until the turn of the twentieth century was sleep accepted by scientists and physicians as being primarily internally driven, regulated by circadian rhythms that take some input from their environment but are principally understood as individual. This led in time to the genetic understanding of sleep and circadian rhythms, with the acceptance that individuals are biologically predisposed to particular forms of sleep and that sleep disorders are hereditary in nature. What I focus on in my discussion of Nathaniel Kleitman and, arguably, his most famous experiment (his descent into Mammoth Cave in 1939) is this turn to the internalization of desire: how sleep is governed by forces that are ultimately beyond our individual control, yet compose us as individuals. In this turn to internalization, individual habits are not what can control sleep and wakefulness, but rather the intimacies integral to the masses are. Sleep is to be attuned to nature and controlled through everyday formations of life and its capacities.

American Sleep, 1690–1937

To return to Mather momentarily, part of his understanding of sleep as avoidance of labor is due to our ability to indulge in sleep, to give way to the carnal desires that sleep represents. In the following, he cites sleep as one of the temptations associated with the devil and his machinations:

> Where shall we find that the Devil has laid out most fatal snares? Truly our snares are on the bed, where it is lawful for us to sleep; at the board, where it is lawful for us to sit; in the cup, where 'tis lawful to drink; and in the shops, where we have lawful business to do. The Devil will decoye us, unto the utmost edge of the liberty that is lawful for us; and then one little push, hurries us into a transgression against the Lord.[4]

Sleep, as early as Mather's colony in Massachusetts, has become associated with death and inefficiency. It is also intimately tied to the everyday, for each day we are subject to the temptations to lie in bed too long, to indulge too much. And at the end of life, we will be judged against those daily indulgences. These same problems were associated with sleep well into the 1800s, when similar moral discourses were used during the early industrial period in an attempt to produce a laboring mass of consolidated sleepers.

Benjamin Franklin, like Mather, was no physician or scientist. But his moral prescriptions on sleep collaborate in the foundations of economic and moral thinking about sleep and wakefulness. In his essay "An Economical Project," published in the *Journal of Paris,* Franklin takes the French to task for their late rising. He writes, teasingly:

> Your readers, who with me have never seen any signs of sunshine before noon . . . will be as much astonished as I was, when they hear of his [the sun's] rising so early; and especially when I ensure them, *that he gives light as soon as he rises.* I am convinced of this. I am certain of my fact. . . . And, having repeated this observation the three following mornings, I found always precisely the same result.[5]

The problem that Franklin identifies among the Parisians, especially the elite among them, is that they spend their nights in revelry and retire to bed too late to take advantage of the sun's light. Instead, because they start their days after noon, they stay up late and burn tens of millions of pounds worth of candles each year in support of this habit. Franklin proposes that they instead wake up with the sun, ending their evenings much earlier and saving the cost of the millions of candles they would otherwise require. He explains, "Oblige a man to rise at four in the morning, and it is more than probable he will go willingly to bed at eight in the evening; and, having had eight hours sleep, he will rise more willingly at four in the morning following."[6] Franklin's recourse to habit—the daily rising at a set hour, leading to the consolidation and timeliness of sleep—is echoed by those who follow him, more properly in the idiom of medicine and science. So too are his assumptions about the economic value and efficiency of sleep, which are returned to repeatedly through recourse to the interconnection of humans with nature and their environment.

Although not the first or the most prominent work to do so, William Whitty Hall's *Sleep* emblematizes the connection of medical and moral thought that is apparent in both Franklin's and Mather's thinking on sleep.[7] Hall's model of medicine and control primarily focuses on individual behaviors and the need to shape them appropriately on the basis of his conception of nature. At the base of his understanding is the acceptance that acting in accord with nature is the most efficient form of behavior. In making this argument, Hall relies on two refrains for support: the need for proper ventilation and the moral propriety of bed sharing. In the case of the former, he aligns himself with the then-dominant view of the causal effects of impure air, which he summarily employs as the basis for moral, prescriptive rules regarding cohabitation. Regarding the latter concern, Hall is responding to the contemporary extension of the family home into a number of rooms, each with individual occupants, formerly (and in large part still) the provenance of the wealthy.[8] Hall's hygienic cosmology is summarized in the following passage:

> It is of the utmost consequence that every practical and rational means for securing a pure air for the chamber should be employed, the most important of these being large rooms and single beds. . . . it is unnatural and degenerative, for one person to pass the night habitually in the same bed or room with another, whatever may be the age, sex, or relationship of the parties. . . . the atmosphere of any ordinary chamber occupied by more than one sleeper, is speedily vitiated, and . . . in this vitiated condition, it is breathed over and over again for the space of the eight hours usually passed in sleep.[9]

Not to be undone by the forces of custom, Hall preempts criticism by noting, "It is not denied that two persons have slept together in the same bed for half a century, and have lived in health to a good old age; this only proves how long some may live in spite of a single bad habit."[10] Throughout *Sleep,* Hall takes "habit" as his object, resolute in his focus under the auspices of a more general hygienic project. In the following passage, Hall argues for the synergistic force of culture and nature, reasoning that it is through "regular" rhythms that a "natural" pattern of sleep might emerge:

> A certain amount of sleep rests, renews, and strengthens the whole man, but to accomplish such a result, sleep must be regular. . . . the general habit should be to retire at the same hour in the early evening of every day. In a short time the result will be an ability to go to sleep within a few moments after retiring, and to sleep continuously until

morning, provided the sleeper leaves his bed at the moment he first wakes up, and does not sleep during the day. . . . Let there be an appointed time, not to be changed for any common reason; the feelings will come at that appointed time, and when satisfied, nature calls for not more until the appointed time comes round again.[11]

Hall tacitly argues for sleep as a flexible inevitability for humans; it must happen, but how and when it happens is up to the will of the individual. The problems with sleep are perceived as cultural ones; while nature is accepted to tend toward a state of balance and harmony, humanity derails such predispositions. This is most evident in Hall's attacks on "second naps," a view that incidentally corroborates the biphasic model of human sleep as a form that existed through the latter half of the nineteenth century.

Second naps, for Hall, are problematic in that they are clearly the choice of humans to override the "appointments" of nature. His solutions for overcoming the desire for second naps are wholly habitual, as evidenced in the following:

> Nature must make the appointment, and will always do it wisely and safely; and there is only one method of doing it. Do not sleep a moment in the day, or if essential do not exceed ten minutes, for this will refresh more than if you sleep an hour, or longer. Go to bed at a regular early hour, not later than ten, and get up as soon as you wake of yourself in the morning. . . . It is not absolutely necessary to get up and dress, but only to avoid a second nap.[12]

Hall provides additional explanation for the need to avoid second naps largely as a remedy for nocturnal emissions (which he refers to in the following as "exhaustions"). Hall argues that to avoid unwanted ejaculation, one must regularize sleep through habitual practices of waking at appointed times. He explains:

> Exhaustions occur in the unsound sleep of the later part of the morning, often during the "second nap," as it is called, it may be added that persons who sleep in the daytime, and thus render the sleep of night less deep, are more troubled with these things. By going to sleep at a regular early hour, say not later than ten o'clock, by not sleeping a moment in the daytime, and by being regularly waked up at the end of seven hours, which is about as much as persons usually

require, the sleep would, generally, in a week be sound, deep, connected and refreshing. . . . nature loves regularity so much she would waken up the body within a few minutes of the times, if only the habit were persistently followed of getting up at the very first moment of waking, or at least, by strong exercise of the will, avoiding a second nap.[13]

In this passage, Hall evidences the problematic aspect of habits, namely, that they must be persistently followed: they are hardly inevitable. Inasmuch as habits may override nature, they have a logic of their own and depend on the willful practice of individuals. This need to persist in habitual action also demonstrates the irregularity of everyday life, because it depends on the willfulness of individuals who may be unpredictable. Thus, the need for instilling a desire for a particular form of sleep becomes central for those who would order everyday life. As is also the case in convincing individuals not to share a bed, Hall rallies whatever support he can find for the purpose of altering sleeping habits in relation to the use of time; otherwise second naps may remain a part of everyday life for those who persist in their avoidance of his prescriptions.

Throughout *Sleep* Hall exploits scientific explanations for why beds should not be shared, including such justifications as the transfer of electricity between bodies, the contamination of air, and the corrupting, almost vampiric influence of the aged lying with the young. But his ultimate explanation is a transcendental natural order, which, above all, strives toward an inevitable rhythm of regularity, moderation, and continuity. He writes:

> The appetites . . . are to be gratified at stated times, and at none others; they are not to be teased or tempted or stimulated by always having at hand the facilities for gratification; these occasions being determined at first by the decided calls of nature, which will then be made regularly, moderately, and continuously, to the end of life.[14]

But the calls of nature can be easily interfered with, both in convenience and abundance. And also, it seemed, by culture itself: Hall's use of "stated time" belies nature's abilities to be fully coercive in its determinations. To employ the need for stated times is to evidence how dependent nature's rhythms are on culture, on the spatiotemporal sense of inevitability that only the ordering of everyday life provides.

More central to nineteenth-century sleep science is the work of William Hammond, who, in his *Sleep and Its Derangements,* provides a watershed of allopathic thought on the subject of sleep. Hammond accepts that sleep is "an inexorable law of his being [human],"[15] that sleep is an irrevocable force of nature that interferes with culture and society and is interfered with by them. He argues:

> Habit exercises great influence, and thus individuals, for instance, who are accustomed to continual loud noises, cannot sleep when the sound is interrupted. . . . the predisposition to sleep is, in healthy persons, generally so great that when it has been long resisted, no sensation, however strong it may be, can withstand its power.[16]

It is clear in both of these quotes that for Hammond, something in the nature of sleep is beyond our human control. He explains this with recourse to mechanistic models, comparing human physiology to steamers and engines. At the base of his comparisons is an understanding of efficiency and labor power as motivating human desires for sleep. Sleep, from the perspective of a nineteenth-century thermodynamic conception of the human body, is intended to restore the body to its most efficient state and to replenish its capacities. Hammond explains this:

> The necessity for sleep is due to the fact that during our waking moments, the formation of the new substance does not go on as rapidly as the decay of the old. . . . The more active the mind the greater the necessity for sleep, just as with a steamer, the greater the number of revolutions the engine makes, the more imperative is the demand for fuel.[17]

In Hammond's thought, there can be substantial differences in the desire for sleep among individuals. For Hall, the habits governing sleep are based on the assumption of the masses: the same rules apply to everyone equally, which results in the individual and the masses sharing the same qualities. But this is not so for Hammond, who sees individual variation of sleep desires as particularly human.

Hammond outlines his understanding of individual needs in this way, marrying understandings of habit, variations within the animal kingdom, and degrees of human intelligence to explain why some individuals need more sleep than others. He writes:

The necessity for sleep is not felt by all organic beings alike. The differences observed are more due to variations in habits, modes of life, and inherent organic dispositions, than to any inequality in the size of the brain, although the latter has been thought by some authors to be the cause. . . . In man, for instance, persons with large heads, as a rule, have large, well-developed brains, and consequently more cerebral action than individuals with small brains. There is accordingly a greater waste of cerebral substance and an increased necessity for repair.[18]

Hammond firmly places human sleep patterns in the realm of the cultural, emphasizing that it is habits and modes of life that take precedence over the size of the brain. His argument that some individuals need more sleep on account of their intelligence hinges on his conception of inherent organic dispositions: the desire for sleep has to do with the *natural* basis of human intelligence, granted in the size of individual brains. This is in juxtaposition to inequality in the size of the brain, which marks the possible variations within kinds of organic beings. Hammond explains this variation within species in the following way, comparing the consumption of the brain's capacity to the functioning of a steam engine:

The immediate cause of sleep is a diminution of the quantity of blood circulating in the vessels of the brain, and . . . the exciting cause of periodic and natural sleep is the necessity which exists that the loss of substance which the brain has undergone, during its state of greatest activity, should be restored. To use the simile of the steam-engine again, the fires are lowered and the operatives go to work to repair damages and put the machine in order for the next day's work.[19]

The belief that our brain's power can be diminished and that some individuals have greater capacity for the use of the brain has its logic in nature. This appeal to the natural is also an appeal to efficiency, as natural sleep is intended to restore the brain's powers and capacities in periodic fashion, leading to the maintenance of life and the continued ability of individuals to labor.

Sleep, for Hammond, as is the case with many of his successors, is troubling because of its apparent ability to occur beyond the will of the individual, this being the other aspect of its status as an inexorable law.

This is apparent in his discussion of two quite different phenomena: the first being the ability to set one's own internal alarm clock, and the other being the experience of insomnia. In the case of the former, Hammond suggests, "The brain is, as it were, wound up like the alarm clock and set to a certain hour. When that hour arrives, an explosion of nervous force takes place, and the individual awakes."[20] This suggests that there are means within human power to control sleep or at least to determine when wakefulness occurs. But compare this with Hammond's description of lying in bed awake, struggling to sleep: "The more the will is brought to bear upon the subject the more rebellious is the brain, and the more it will not be forced by such means into a state of quietude."[21] Such a conception of the will and the brain pitted against each other both accepts the inevitable powers of nature and posits them in opposition to the domain of intention, agency, and habit. That sleep exists beyond our human controls has a more profound implication, namely, in the attribution of culpability in the actions that occur in our sleep. Because of the inexorable powers of sleep and our limited ability to control it, Hammond suggests, "During sleep the power of bringing the judgment into action is suspended. . . . we cannot exert the faculty of judgment in accordance with the principles of truth and of correct reasoning."[22] Hammond's successors embrace the conception of sleep as restorative, efficient, and tied to natural rhythms, and they leave this question of intention aside.

Writing thirty years after Hammond, Andrew Wilson continues the use of mechanistic metaphors for understanding the functions of sleep and its powers, in his article "The Ape of Death," a state of the science piece written for *Harper's*. But what infuses Wilson's language is more intense economic language and metaphors. Take his description of the process of sleep and its purpose: "Sleep . . . would be determined by the fatigue of the brain cells and by the contraction of their processes, while waking would be the result of the renewal of business in the busy cerebral exchange, when the cells, revived by their rest, make contact with other cells."[23] At the end of the nineteenth century, "business" and "the busy cerebral exchange" expose how economic the control of sleep has become; it saturates scientific understandings of sleep and the desires of individuals. But Wilson employs mechanistic metaphors as well, now in terms that are explicitly related to labor. He writes of the heart, "[The heart] is in the position of a workman who takes short intervals of rest betwixt short spells of work. . . . It is really inconceivable that any vital structure, large or small, should labor incessantly."[24] This conception of labor and rest, which even the heart undergoes, lays the foundations for Wilson's understanding of human physiology, as well as for society. The implication of such a mechanistic conception of human everyday life and sleep is that there are more and less efficient ways to order the former and desire the latter.

Marrying conceptions of society and physiology, Wilson suggests:

> The alteration of rest and work is by no means limited to our own existences, nor, indeed, to the brain itself as the organ chiefly concerned in the oscillation between labor and repose. . . . The living engine slows down, as it were, and banks up its fires so that its pulsations are sufficient, not for actual labor, but for merely maintaining the passive flow of force within the organism.[25]

Labor and the living engine structure our periods of rest and work. The prominent figure of work, not life, in Wilson's description of human sleep and its purpose lays the foundation for his understanding that the body must recuperate, doing away with waste so that the brain and body can return to their full labor capacity. Wilson argues, "Brain-labor, equally with that of the body at large, obeys the inevitable law that all work means waste, and therefore the waste products of the brain cells must assuredly be removed as pure blood must be supplied to regenerate and renew their energies."[26] Sleep, by the time of Wilson's writing at the end of the nineteenth century, had become intimately tied to conceptions of labor, relating to both the body's capacity for productive action and also everyday bodily being in the world. This echoes the work of early twentieth-century sleep physicians and scientists, particularly in efforts to popularize the need for sleep and its management.

In a short book published by the National Health Council, Donald Laird both works to popularize contemporary conceptions of human sleep and provides prosaic advice, captured in the book's title: *How to Sleep and Rest Better*.[27] As in the theories of many of his forebears, efficiency is central to Laird's conception of sleep—how to sleep productively to make the most of the day. Also like his intellectual precursors, Laird uses mechanical metaphors for conveying how sleep functions for the human body. In the following, he compares the impacts of lost sleep to driving an automobile in the wrong gear—possible, but inefficient:

> After losing only a slight amount of sleep we can do about the same grade of such mental work [as simple arithmetic], but it takes more out of our bodies to do it.
>
> Compare this with driving an automobile. Using additional fuel, any car can be driven in a low gear. The same distance will be covered as at high speed, and the added cost of gasoline will not be enormous. But the car does not get to the finish mark nearly so quickly and the engine will be in poor shape when a few thousand miles have registered on the speedometer.[28]

To approach the day from the position of an inadequate amount of sleep is to expend more of our capacity in tasks that might otherwise require very little effort. To offset this, one needs to attend to the indicators that one's sleep is inefficient or abnormal. In his discussion of "signs of poor sleep" Laird lists, among others: "Waking up before obtaining as much sleep as the average. Difficulty in waking up after having as much sleep as the average for one's age. . . . Tired or inactive when awakening from a normal amount of sleep."[29] Whereas thinkers like Hall and Benjamin could make reference to what they conjectured to be normal or average sleep, by Laird's time the consolidated model of human sleep was fairly well established. This provides Laird with a normative basis for his claims in terms of what sleep should be but also its variations and functions.

At the heart of Laird's assumption of inefficiency is our inability to relax when the opportunity to sleep arises. Our desires and intimacies, the interactions that compose our everyday lives, stimulate us to the point of distraction, even when lying in bed. Throughout *How to Sleep and Rest Better*, Laird extols the virtues of relaxation techniques, explaining, "Your efficient sleeper . . . has acquired this trick of relaxing in one way or another. True, some of us relax completely and naturally, as in healthy childhood, all through life."[30] Laird recommends breathing exercises coupled with the tensing and relaxation of muscle groups, a somatic practice to relax into sleepiness. But the intelligent are prone to be nervous, to be insomniacs. His points of comparison, not unlike Hammond, are "the feeble-minded," who are prone to easier sleep. Laird writes:

> Feeble-minded folk sleep better than brainy ones because their
> minds are habitually empty; the moron does not need to make his
> mind a blank before going to bed since it is blank both day and
> night. One price that is paid for brains is troubled sleep, and altho
> the feeble-minded sleep most, it is the person with brains who needs
> the best sleep.[31]

For those with brains who might have difficulty settling into sleep, or not
even want to sleep for fear that it cuts into their labor potential, Laird has
another recommendation to maximize our nightly sleep:

> To make our sleep constructive we should review in our minds near
> the retiring hour some of the things we want to do on the coming
> day, so that our so-called subconscious processes can be active on
> these things to be accomplished rather than with disturbing dreams.
> Taking our tasks to bed for solution helps the tasks, and also helps
> sleep if we are otherwise emotionally calm. The person who keeps his
> mind at work while asleep does not have his sleep disturbed—if his
> emotions are stable and if he is facing life with a smile.[32]

Good sleep for Laird is efficient in two ways: first, it ensures that we are
at our greatest capacity throughout the day, providing us with the ability
to attend to our many tasks. Second, it can serve to aid our daily tasks
by helping us to process them. Although Laird is interested in helping
his readers sleep and rest better, ultimately *How to Sleep and Rest Better*
exhibits a modest dismissal of the importance of sleep; it would be better, Laird argues, to spend our sleep with our subconscious processing in
anticipation of the next day than to relax in dreaming. Spending a third
of one's life in sleep is hardly an efficient use of one's time!

How to Sleep and Rest Better ends with a questionnaire for its reader.
The purpose of this twenty-question self-exam is to determine one's "sleep
unit percentage."[33] How efficient is your sleep, and how might it be more
so? Laird asks his readers to consider questions like the following: "Do
you go to bed with the idea that you are going to sleep? Do you refuse to
court the idea that you won't sleep?" Too many negative answers to these
questions add up to inefficient sleep. This is the inevitable end of the
moral and economic thinking that has guided American conceptions of
sleep since the seventeenth century. Sleep's continued conceptualization
as tied to labor and efficiency is indebted to widespread interests that find

their roots in early American thought. These same concerns continue to lay the foundations for scientific and medical conceptions of sleep but are changed by the work of leaders in twentieth-century sleep science such as Nathaniel Kleitman and William Dement.

Isolating Sleep in the Twentieth Century

In 1938, by descending into Mammoth Cave with a graduate student as a fellow test subject, Nathaniel Kleitman, professor of physiology at the University of Chicago, managed to interest the American public in the science of sleep and its potentials.[34] Kleitman's plan was to synchronize himself and graduate student Bruce H. Richardson to a six-day week, with each day made up of twenty-eight hours. They would remain in the cave for thirty-two days to explore the changes made in their circadian rhythms (a concept only then gaining some credence), primarily ascertained through variations in their body temperature. It was Kleitman's hope to conduct this experiment in relative quiet, but when the press discovered his plans, the experiment became the subject of national attention. The test, as recounted by Kleitman, was an ambiguous success at best. With only two test subjects, generalizing the results of the test would have been difficult regardless of its outcome; of the two subjects, only Richardson's circadian clock was able to be entrained to the new schedule, which might easily be attributed to the general sleep deficits of students rather than the flexibility of circadian rhythms. Kleitman, meanwhile, had great difficulty with the new schedule and was unable to acclimate to it fully. Unwilling to nap, Kleitman found himself drowsy throughout the "day" and unable to sleep during the "nights" he had scheduled for the experiment. My interests in recounting this experiment here is in how Kleitman set limits on what sleep can be and employed these expectations in later experiments, thereby concretizing it. Modern attitudes about sleep—about productivity and efficiency—find justification in Kleitman's understanding of natural sleep, especially as these attitudes relate to habit. In this section, I first focus on Kleitman's experiment in Mammoth Cave and then turn to his later consultancy with the U.S. military regarding the tactical use of normative sleep. Normalizing sleep through scientific practice led to the eventual development of clinical understandings of pathological sleep.

The goal of moving the experiment into the cave, as it was expressed by Kleitman, was primarily and specifically related to the environmental factors that control sleep, especially daylight and noise. Secondarily, moving into the cave meant extricating both Kleitman and Richardson from their social obligations. As described in a press release from the University of Chicago's Department of Press Relations on July 8, 1938 (immediately following the return of Kleitman and Richardson to the surface world):

> Under the uniform conditions of temperature, illumination, and quiet of the cave, [Kleitman] and Mr. Richardson sought to adapt themselves to a 28-hour daily cycle, or a six-day week. . . . Free from external influences such as sunlight, activity of others, and temperature changes, Dr. Kleitman and Mr. Richardson sought to change their *habits* while in Mammoth Cave to a twenty-eight hour day. Dr. Kleitman's temperature curve tended to remain that which prevailed during a normal twenty-four hour day, but Mr. Richardson's curve was readjusted to the longer cycle.[35]

The press release quoted Kleitman as saying, "This experiment merely confirms previous results which I have obtained, in demonstrating that the cycle is dependent on the activity of the individual. Some individuals can change their cycle with considerable ease; others find a cyclical change difficult to establish." In making such a claim, Kleitman was accepting the possibilities of human variability and that sleep is at once flexible and inevitable: Sleep can be governed by individuals, willfully or physiologically. It is subject to habits. It should be noted, however, that Kleitman disallowed napping in both the Mammoth Cave experiment and previous experiments, despite what desires for sleep test subjects might be experiencing; he was already convinced of a consolidated sleeping pattern for humans, and his experiments reified this expectation. Individual variation, yes, but within the confines of set norms.

Kleitman follows his intellectual forebears in tracing the relationship between "habits" and biology that they established, which he builds into his experimental design. What unites the space of Mammoth Cave (and later, sleep science more generally) and American society is the conversion of individual experiences of sleep into the sleep of the masses: the experiences of individuals are taken to be representative of society

Kleitman's schedule in the other armed forces; Kleitman closed his overview of the schedule with the following appeal to military rationale: "A SLEEPY FIGHTER IS A MENACE TO HIMSELF AND HIS COMRADES, BUT NOT TO THE ENEMY."[39] Building on his work in Mammoth Cave, Kleitman worked to establish the need for ordering society—the military in this case—on the basis of efficient uses of human sleep. This efficiency was founded in physiological predispositions and also in arbitrary ordering.

On the basis of his experiences as an observer on the USS *Dogfish* in 1948, Kleitman was called upon by the U.S. Navy in the summer of 1949 to develop a new schedule for submarine crews. Crews had traditionally worked in "four on—eight off" schedules, meaning that they stood watch for four continuous hours, took an eight-hour break, and then returned to post for another four hours, before another eight-hour period of recreation. Because the eight-hour sleep requirement expected by officers and crew was scheduled so tightly between the work shifts, the crew was often awakened prior to having sufficient sleep; the result of this was that napping occurred throughout the other relaxation period, which was perceived as a problem by Kleitman. His experimental schedule broke the crew into three sections, with each section working a total of eight hours per day, but their shifts were broken into two three-hour periods and one two-hour period in an attempt to stave off potential boredom or fatigue while at post. Moreover, each section of the crew was also given a longer period of unbroken free time, ranging between ten and twelve hours, which would ideally rotate among the various shifts. As Kleitman states in his proposal, "An unbroken stretch of free time . . . permits one to sleep 8 to 10 hours and still get up at least 1½ hours before the beginning of the first watch period. Naps during the 2 and 3 hour intervals between watches are unnecessary. Though permitted, they should be discouraged." Why naps should be discouraged, however, was left vague. He went on to explain, "All meals served within 10 to 12 hours, as customary, but in the usual order of succession, breakfast always coming after the 'long' sleep. . . . Breakfast served one hour before the beginning of the first watch, allowing a 'morning' waking-up period."[40] Other meals also kept to cultural norms, with supper being served three and a half hours before bedtime, allowing, tacitly, for a period of informal recreation after being served. Central to Kleitman's experimentation were cultural expectations of what the everyday should be and the assumption that science was intended to preserve it, even when at sea and in the absence of familiar cues and social obligations that are usually required to ensure that the organization of everyday life makes sense.

The experiment was a disaster, in part because the submarine crews put the new schedule into effect for only eleven days. Kleitman suggested that in the future, if such experiments were to be conducted, a "responsible investigator" should be aboard to control the experiment and ensure its success, rather than allowing the crew and events to conspire and derail proper scientific protocol. The summary of the surveys given to the crew aboard the USS *Tusk,* prepared by R. K. R. Worthington, complained of the lack of space for the crew to inhabit during their recreation period; some of the crew inhabited the bunk rooms, others were at stations, and the remainder were left adrift, "loafing around trying to keep out of the way." As a result, Worthington implied, the men stayed in the mess hall and ended up eating more than three meals per day, because "they were up and had little else to do." Eating extra meals might not normally be a problem, but it was when at sea and under wartime conditions with limited rations. Worthington claimed the cramped space of the submarine was at fault for the experiment's failure, citing "the operational and structural conditions imposed" as necessarily limiting the possibilities for temporal rearrangements of life underwater. Naps, prior to Kleitman's best efforts, were substantial, with 91 of the 196 crewmen claiming to take naps of at least two hours per day; only 46 took naps less than an hour long, and 59 crewmen took naps longer than two hours. During the experiment, 81 of the 173 surveyed crewmen claimed to take no naps, 64 took up to two naps daily, and 28 failed to answer the question. Only 33 crewmen claimed the new sleep pattern was "better," with 98 claiming it to be "poorer." A general lack of drowsiness was also reported, with 76 of the crew claiming to experience it "not at all" and another 68 claiming that it still recurred "at times." In response to the question "Conducive to greater alertness?" 117 crewmen replied in the negative, and 143 of the crew argued that the new schedule should not replace the traditional one. Despite Kleitman's best efforts, the navy crewmen were resolute in not wanting the temporal regimes of normative everyday American life at sea. These experiments were conducted again in 1952 onboard the USS *Scabbardfish,* with 48 percent of the crew supporting the alteration to the experimental schedule. Despite this, the U.S. Navy preferred, it seems, a life of naps and "abnormal" uses of time and rejected Kleitman's attempts to reorder life at sea yet again.

In 1955, by invitation, Kleitman participated in the Subcommittee on Field Sleeping as part of the Environmental Protection Division of the Quartermaster Research and Development Center; he served as a critic of the research conducted on the psychological, physiological, biophysical, and psychophysiological aspects of field sleeping. In his response to the various findings of the subcommittee, dated April 22, 1955, which suggested that the "definition of sleep, the phenomenon we wish to measure" still remained obscure to scientific knowledge, Kleitman argued, "By comparison, one does not tackle a time–motion problem of the assembly line by first trying to find adequate definitions for 'time' and 'motion,' but rather by studying patterns of performance." He went on to argue, "'The nature of sleep,' fascinating and intriguing though it may be, is no a problem for the Quartermaster"; rather, the point was "to test the quality of the sleep by the subjective response of the sleeper and, even more important, by the performance of the subject during the waking hours that follow." He recommended, "Critical performance tests should be made in the field, rather than in the laboratory, on service men, rather than on students or hired subjects, on performance on the shooting range, rather than in mental arithmetic, and in the middle of the day rather than immediately upon awakening in the morning. All research projects proposed should be evaluated in the terms of promise to furnish an answer to the criteria of a 'good night's sleep.'" What Kleitman was foundationally interested in, and what he attempted to steer the subcommittee toward, was a strictly biological understanding of sleep and its effects, in as warlike conditions as could be manufactured, not in the laboratory per se, but under controlled conditions nonetheless. Only through this careful observation and control could efficient uses of sleep be ascertained. This could be achieved through the management of biology in social contexts rather than through pharmaceuticals or other technological means, as Kleitman suggested in his critique:

> Attention should be called to the diurnal variation in *efficiency* of performance resulting from the acculturation of human beings to the daily schedule of family and community living. The daily rhythm, as revealed in the diurnal body temperature curve, can be either fortified, where activity is restricted to regular hours, or appropriately

modified where the group performance is on a round the clock basis (as on ships, radar towers, hospitals, transportation, communication, sentry duty, etc.), with necessary adjustments in *the hours of sleep*. (emphasis added)[41]

Kleitman's attempts to increase military efficiency and garner tactical advantages for the American military were marked by a number of scientific and cultural expectations. For Kleitman, something akin to normative everyday life was worth preserving under wartime conditions with its diurnal rhythms, because it was seen as efficient; despite this, his primary means of achieving the advantages he sought was through the management of daily temporal regimes. Only at the turn of the twenty-first century would such pursuits be seriously considered, and not through social manipulation of sleeping times, but largely focused on biological means. Everyday life might be rearranged, but this was necessary only when in competition with another society to garner a tactical advantage. This might be the case not only militarily but also economically: the rise of the twenty-four-hour society is a similar effort to outstrip competitors who would happily sleep through the night. The everyday, for Kleitman, might be tinkered with, but its consolidation needs to be upheld. For the sake of efficiency, it is vital that the everyday hews closely to the natural, diurnal cycle. In this context sleep medicine would be developed, emphasizing consolidation of sleep and everyday life for the sake of efficiency.

Sleep Now: William Dement and Modern Life

William Dement, as the primary steward of sleep medicine through the later half of the twentieth century, worked to develop a paradigmatic understanding of sleep as a natural inevitability: something that must occur but can be varied slightly by individuals and controlled through individual habitual practices. Habit works to align individuals with the stable organization of everyday life. In this context, medicine serves as an intermediary between the biological demands of the individual and the expectations and obligations of society. Disorders of sleep, in this paradigm, are in need of alignment with these forces (the social and the biological, the everyday and the individual), although most often social obligations prove more persuasive, forcing the biological to be altered through its flexible capacity to meet social order. This persuasiveness of the social results from the ability of economic interests to make demands on individuals to meet obligations, through their desire for everyday life; the temporal and spatial fixity of such social demands, paired with a unified understanding of the human phenomenon of sleep, provided twentieth-century sleep medicine with markers against which to measure individual sleep disorders, while also providing concrete understandings of what normal sleep is. Dement perceives sleep as primarily a natural, inevitable force, indebted to the habits of individuals, but a force that is culturally flexible, as evidenced in cultures that practice midday naps[42]—a view not wholly dissimilar from that of his predecessors in the nineteenth century. These midday naps are viewed by Dement as a variance from the biological norm of eight hours of consolidated sleep, rather than as an alternative or more likely possibility for human sleep patterns; in other words, naps are understood by Dement as a culturally influenced choice rather than a biological necessity. From this foundation, one might understand the need to nap during the day as produced by the choice to stay up late the night before in resistance to the spatiotemporal order of society, a choice facilitated by the advent of electric light and technological distractions. These choices interfere with the efficiency of sleep and daily life and lead to the development of "sleep debt," mounting lost sleep on the part of individuals and society more generally.

Dement's critique of contemporary American life and its technological abuses depends on the construction of a state of nature from which a fall from grace has occurred. Presumably, living in accord with nature also leads to an efficient life. He argues:

> In virtually every aspect of contemporary living—from electric lights to all-night television to split shifts at work—we are literally punching the clock that maintains the synchronicity of our mind and body. In just a few decades of technological innovation we have managed totally to overthrow our magnificently evolved biological clocks and the complex biorhythms they regulate.[43]

Moreover, he claims, "Our loss of sleep time and natural sleep rhythms is the tragic legacy of a single and profound technological advance— the light bulb. . . . Edison accomplished something Prometheus could not imagine, because he separated the light from the fire and offered it for our infinitely more convenient and *flexible* use" (emphasis added).[44] Dement is not alone in telling this apocryphal tale: the "state of nature" wherein humans lived in synchronized rhythm with an agrarian nature is a persistent story and has been employed by numerous actors to legitimate the contemporary spatiotemporal regimes of American everyday life.[45] What is particular about Dement's understanding of the changes that are wrought by the advent of cheap electrical lighting is the way it allows culture to supersede nature: the need for sleep and the chronicity of life on Earth is increasingly mediated by the will of the individual who can break from these biological and geological predispositions and form a different, and potentially less efficient, desire for sleep. Not only is it a decision that individuals can make, but it is also subject to the whims of societies themselves, as some may choose to extend their days with "split shifts" and "all-night television" and others may not. The flexibility that electric light allows for individuals and societies, the changes in sleep that this produces, stands in opposition to an inevitable natural order to be understood as a causal explanation for mounting sleep debt in American society. There is the efficiency of nature on one hand and the attempts at efficiency on the part of society on the other; the latter fails in the face of the former.

Throughout *The Promise of Sleep,* Dement makes recourse to what sleep was like previous to the demands of modern life, arguing pointedly, "Once we use electric lights, our [biological] clocks start lagging about an hour every day."[46] This is understood as a result of the progressive push of human circadian rhythms, which are accepted to exceed the twenty-four-hour daily clock by upward of an hour in some experimental settings. Electric lights allow individuals to lengthen the day, and the presence of sustained bright light positively reinforces the desire to stay up later each night. Dement can hold such views about the lengthening of the day and the destructive potential of electric light only by suppressing the alternative forms of sleep and electric light's antecedents, which have all interfered with biology and natural order. Echoing Hall, Dement argues, "Most likely we need the sleep debt accumulated during our waking 16 hours, plus a little extra, in order to fall asleep in 5 or 10 minutes and sleep through the night. The idea that a little sleep debt is good is a revolutionary concept."[47] In such a view, the progressive quality of the circadian rhythm becomes necessary for sound sleep, although because of the nature of fixed spatiotemporal orders of everyday life, the amount of sleep an individual gets in a given night decreases through the week, as one's circadian cues for sleep progress later into the night, while the timing of social obligations remains constant. Thus, Saturday mornings have become a time to make up for a week's worth of diminishing sleep, which begins again on Monday, resulting again, ideally, in five nights of sound but diminishing sleep throughout the week, rather than five nights of short, variable sleep supplemented by daily naps.

The conception of sleep that Dement forwards is one that is intimately tied to nature but lends individuals the ability to make destructive decisions, which largely work to divorce humanity further from a natural state. Dement posits this ability as a specific capacity of primates, arguing, "Primates, including man, are able to compress their daily need to sleep into eight hours because they sleep more deeply and much more continuously than if there were no daily period of sustained wakefulness."[48] Despite what might be read as an evolutionary hypothesis regarding humanity's potential for shaping its spatiotemporal orders, both at the individual and broader social levels, very little attention is paid on Dement's part to the potential relationships between sleep disorders and evolution. This is particularly peculiar in that Dement understands humanity as innately linked with its environment, noting, "It is Earth itself that must act as a metronome, a timekeeper setting the tempo of our days. The bright light of morning and its dimming at dusk must synchronize our clocks each day, calling us awake and lulling us to sleep."[49] In this view, in which humanity's rhythms of sleep are produced by the Earth but there is the possibility of altering these rhythms through choice, sleep disorders are potentially understood as both biologically determined aberrations and socially produced decisions, as both disorders of physiology and desire. This results in two complementary forms of treatment, one that focuses upon fixing the behavior of the patient, the other that often promotes surgical, prosthetic, or pharmaceutical options. Which of these treatments is employed depends not only on the disorder but also on the way the disorder is understood by the attending clinician. This is precisely what Dement attempts to curtail in the second half of *The Promise of Sleep,* which is a voluminous study of the various sleep disorders that he presents in an effort to operationalize his conception of sleep in the clinic.

Dement works to categorize the various understandings of sleep disorders into unitary nosologic, but differentiated, phenomena. In this way, the many sleep disorders—insomnia, restless legs syndrome, narcolepsy, sleep apnea, delayed sleep phase syndrome, and so forth—are all understood as resting upon a biological foundation of eight quiet, motionless, and consolidated hours of sleep, positioned at an environmentally conditioned time, between sunset and sunrise, but potentially disrupted by individual choices or social conditions. For instance, insomnia and delayed sleep phase syndrome are both understood by Dement as largely the result of cultural or social influences, the former often caused by anxiety or depression, the latter by choices to retire to bed late. In the case of the sleep disorders with recognized biological causes—narcolepsy, sleep apnea, restless legs syndrome—the cessation of each depends upon novel medical treatments. The work of *The Promise of Sleep* is to lay a new clinical and scientific foundation for the sleeping public of the late twentieth and early twenty-first centuries, and it promotes a model of nature and human biology from which all variations are disorders.

One problem for Dement's conception of sleep is that individuals can *choose* to be nocturnal; they can choose to invest in other desires than sleep. Individuals can override natural rhythms of life with technological assistance, thereby taking advantage of the variability of sleep's flexibility. In Dement's words, "Too many adults behave like children when it comes to regulating their own sleep habits, even when their chronic daytime fatigue tells them that something's wrong."[50] In the wake of such behavior, what becomes necessary, in Dement's view, is to construct sleep as something that can be easily manipulated by individuals, but the harsh inevitability of nature will have the ultimate revenge; he argues, "When we fail to acknowledge sleep's sovereignty over our lives or fail to keep sleep healthy, it has the power to kill."[51] Most interesting about this formulation is the positing of an agentive quality for sleep—a conception of the individual in which the body can be at odds with the will—not unlike Hammond's conception of sleep and willpower from a century earlier.

Dement's postulation of sleep and the sleeping public contains uncertainties; these take the form of scientific doubts and social fears. As discussed above, Dement understands sleep as a flexible inevitability, as a choice that individuals can make in relation to their biological demands. Individuals and societies may choose to ignore the need for sleep, but this, as Dement argues, can potentially be life threatening, in addition to simply being inefficient. Individuals and societies may choose to extend days into nights, but this choice can invite sleep disorders, insofar as the sleep needs of daysleepers are understood as variances from social norms of sleep. These decisions regarding sleep and the lengthening of the day through the use of technological means are what lead to those sleep disorders that are primarily understood as socially constructed or constituted through social interactions. For those sleep disorders that are taken to be primarily biological nuisances—namely, restless legs syndrome, narcolepsy, REM behavior disorder, and the various iterations of sleep apnea—medicine can be relied upon to intervene in order to correct the wrongs of evolution or physiology. While these views have developed over time, they are similar to medical understandings of sleep's disorders in the nineteenth century; what is different in Dement's formulation are the ways that medicine is seen as able to treat these disorders. For some nineteenth-century physicians, insomnia and aberrant bedtimes were taken to be social and cultural decisions of individuals; what they understood as "drowsiness" (which would become narcolepsy) was a problem of the nervous system, which they defended against dominant understandings of narcoleptics as being willfully lazy.

The insinuation of this model of sleep into society more generally depended upon Dement and other sleep researchers exporting the conditions of the laboratory to society. The ability to accomplish that depended in turn on an attempt to isolate the laboratory from the social, as in the case of Kleitman's descent into Mammoth Cave: science's results needed to be seen as unbiased. Kleitman and his followers suggested what sleep's nature is, and this nature was returned to American society in the form of a clinical basis for medicine. But these scientific and medical conceptions have come to circulate in society precisely because they accord with dominant understandings of sleep. And, as Kleitman and Dement make evident, their scientific conceptions of sleep are indebted to long-standing cultural traditions of sleep's interpretations. Sleep, as it was understood by researchers and physicians throughout the twentieth century, was not purified of its cultural biases, and yet these biased understandings of sleep laid the foundation for the production of biological norms and their disorders. In so doing, they served both as a means for conceiving of sleep—what normal sleep is and how sleep can be harnessed for the production of life's everyday rhythms on the level of society—and a means to curtail individual habits of patients and other interested sleepers. The agentive power of sleep, that sleep has power over the lives of individuals, however, is what differentiated sleep's surge to public awareness in the nineteenth and twentieth centuries. In the case of the earlier century, sleep was seen as a benign aspect of human life that proper individual habits or institutionalized projects of disciplinary control could direct. In the case of the twentieth century, sleep's nature was increasingly viewed as an unruly one, and while proper habits were seen to have possibly helped to ease complaints, such habits were articulated against an institutionalized use of time and space that depends on the spatiotemporal ordering of sleep for the American population, crystallized around the standard workday, school day, and rhythms of daily family life.

Restless legs syndrome, narcolepsy, and REM behavior disorder are all problematic at the turn of the twenty-first century because of what they index *socially*. Each of these disorders is often construed as a symptom of sublimated intents on the part of individuals rather than as a physiological disorder. The increasing depsychologization of these sleep disorders, as rendered by Dement and his successors in American sleep medicine, consolidated the sovereign power of sleep as a biological function. The idea of the sleeping public, as Dement constructed them and as the idea was put into play by such institutions as the National Sleep Foundation, depends on the acceptance of sleep's powers being biologically inevitable. And variation from sleep's consolidation and hegemony is not simply resistance but also disorder. Allopathic medicine—which turned away from the view of sleep's nature being exposed only in isolation, as practiced by Kleitman—increasingly took as its object the health not simply of individuals but of whole societies, of the masses. Increasingly, then, the foundations of society are conceived of as lying precariously upon sleeping bodies and their disruptions. To order society, medicine must first order bodies to meet society's spatiotemporal demands.

And this is the primary difference between sleep's emergence in the nineteenth and twentieth centuries in the United States: In the nineteenth century, sleep's sovereignty was to be worked with on the part of the individual; through habitual practices one might allay whatever sleeping complaints he or she had. In the twentieth century, particularly in the latter half of it and the beginning of the twenty-first, sleep's sovereignty is met with an equal power, that of pharmaceuticals, of therapies. Some behavioral treatments are used clinically, especially in the case of insomnia and delayed sleep phase syndrome, but pharmaceutical treatments are increasingly relied upon to eradicate sleep complaints. Such dependencies on the efficacy of medicine reify sleep's natural inevitability; only the concreteness and efficiency of pharmaceuticals and their powers derived from nature are effective against such a force. In this reification, which depends on the solidification of sleep patterns affected by Kleitman and Dement, sleep has become not simply a foundation for the production of everyday life but also a potential enemy: if sleep disrupts daily life, then it must be controlled. The flexibility of sleep must be aligned with the variations allowed in the spatiotemporal rhythms of everyday life, rather than everyday life being rendered more flexible for the sake of sleep. In other words, and by way of example, the escalating weekly sleep debt that many 9 to 5 workers in the United States complained of at the turn of the twenty-first century, which is assuaged every weekend by sleeping in, can instead be curtailed through the employment of daily naps. But caffeine and medication, in the service of cultural expectations of normal sleep, suppress the urge to nap throughout the day, and individuals are left to scavenge what sleep they can from a night's rest. This predicament is the result of nearly two hundred years of scientific, moral, and cultural developments in the United States, which have simultaneously resulted in empowering medical practitioners, pharmaceuticals, and caffeine as mediators in individuals' relationships with sleep; rather than a gentle sovereign, sleep has become demonized and rendered an object of medical and scientific control in an attempt to dissolve biological doubts and consolidate the powers of medicine over nature and its efficient management.

3

·················

Sleeping and Not Sleeping in the Clinic
How Medicine Is Remaking Biology and Society

ONE OF THE MOST CURIOUS CASES I WITNESSED AT THE MSDC—not for the symptoms, but for the solutions—was that of an eleven-year-old boy named Ted. The presenting physician described the boy as evidencing insomnia and excessive daytime sleepiness, the latter affecting Ted only between 6:30 and 10:30 A.M. When left to his own devices, the boy would sleep until 10:30 and be fine for the rest of the day. But when he had to go to school, his parents would wake him at 6:30, and he struggled, often unsuccessfully, to stay awake throughout the morning. While at school, rather than simply letting the boy take a nap, his teacher forced him to ride a stationary exercise bike to keep awake. When he would stop pedaling, his teacher would know to wake him or at least call on him. The physicians agreed that the cause was a "clock problem," that is, a dyssynchrony between the child's sleep requirements and the timing of his sleep on one hand and the social obligations of school on the other. Other than an occasional little snore, which the attending physician punctuated with a slight snort-snore, the boy was considered totally physiologically normal. The physicians all agreed that the exercise bike treatment was inappropriate. Given the family situation of the child, reorganizing his waking time and school attendance was impossible. What the family embarked upon was the use of set times for bed and a lightbox in the morning. An earlier bedtime to get his desired amount of sleep and the exposure to bright light upon awakening helped Ted get through his mornings at school. If he were to slip from his schedule and fail to use the lightbox in the morning, his "clock problem" might reassert itself. Similarly, without the demands of his parents and school, Ted's sleeping might again extend until 10:30 A.M. Ted's situation may seem extreme, or comedic, but it offers a glimpse into the many forces that make up everyday life and the management of sleep.

In this chapter, I am interested in the ways that the abstract and theoretical content of sleep medicine is applied to a body of individuals to explain their disordered sleep. What might seem at first glance a straightforward and unproblematic process of aligning symptoms with nosologic categories is, upon closer scrutiny, about more than individuals and disorders. Medical diagnosis is always also an effort at prognostication, of augury: allopathic medicine produces particular futures that depend on rhythmic therapeutic interventions. One rhythm, the disordered, is replaced with an orderly one. This ordered rhythm is produced through pharmaceutical, prosthetic, or behavioral intervention, or any combination of these. Through these means, medical professionals align the everyday lives of individuals with the spatiotemporal demands of the institutions that patients interact with. In the following, I focus on a series of cases addressed by the physicians with whom I conducted my primary fieldwork. Throughout the cases, what I am interested in is how physicians identify disordered rhythms, what they pose as solutions, and how these remedies often, but not always, take as their site of intervention the individual physiologies of patients. Lurking in many of the discussions of cases is the possibility of altering the institutional obligations of individuals and thereby mollifying their symptoms. Before I turn to these clinical discussions, I first examine the increasing popular media representations of sleep, particularly that of the National Sleep Foundation. Sleep is being rendered abstract: it is becoming a means through which to understand a wide variety of social and biological symptoms.

These concerns of media and abstraction, spatiotemporal ordering, and the clinical practice of allopathic medicine are in the service of a more fundamental question: How does the expanding horizon of medicine and its applications allow for both increased medicalization of phenomena once accepted as natural and a fundamental need for "difference" in allopathic medicine? In this chapter, I pursue three interlocked theoretical concepts to address this question: normative ideals, therapeutic normalcy, and integral medicine. Allopathic medicine depends on two fundamental differences: that between the pathological and the normal and that between cause and cure. By focusing first on the media of sleep disorders and its powers of persuasion and then on the clinical negotiations of sleep disorders, I show how allopathic practice is producing and employing difference and sameness as foundational to medical interventions in individual lives. My interests are in clarifying a lens through which to assess the powers of contemporary medicine in the United States and in demonstrating how sleep is but one of the many terrains that medicine produces as a site of action. The emergence of sleep as a site for medical concern has to do with the rise of particular understandings of sleep and its disorders that have occurred since the 1960s, as well as long-held cultural expectations of sleep in the United States, in medicine and in society more generally. Through normative models of sleep, individuals come to identify themselves with society through their individual sleep and its disorders.

Abstracting Sleep and Its Disorders

The National Sleep Foundation (NSF) takes as its central concern "waking America to the importance of sleep."® In order to do this, the foundation has used a wide variety of media tactics, from pamphlets and comic books distributed in sleep clinics to national lobbying days in Washington, D.C.—National Sleep Awareness Week, which culminates with the shift to daylight-saving time in the spring. The most public face of the NSF is its Web site, which has developed over the years from a skeletal outline of information about sleep and its disorders to a rather robust resource on the science and medicine of sleep. For the interested, the NSF also produces a weekly e-mail newsletter. The newsletters summarize findings published in medical journals and, in so doing, attempt to popularize findings for a broad audience who might not understand the technical findings of laboratory scientists, medical researchers, and epidemiologists. Beyond the alteration of language from scientifically precise diction to everyday American English, this process of translation also requires the abstracting of data. In the process of abstraction, precise, situated medical and scientific knowledge is broadened in its applicability. From one population, information is broadened to be applicable to much larger constituencies. Similarly, in the NSF's comic books, intended for children, general assumptions about sleep and everyday life are rendered abstractly applicable to children to understand their sleep and social lives through them. In this section, I focus on these two media, the weekly sleep updates and the NSF comic book *Time to Sleep, with P.J. Bear,* to evidence what this abstraction looks like and how it might be interpreted by those who consume the media.

The NSF's position on sleep deprivation is neatly summed up in the following passage, lifted from one of its weekly alerts: "Too little sleep results in daytime sleepiness, increased accidents, problems concentrating, poor performance on the job and in school and possibly increased sickness and weight gain."[1] How, it might be asked, can one afford not to take sleep seriously when it impinges on so many aspects of everyday life? More recently, the NSF has adopted less alarmist language in attempting to appeal to Americans. Now the NSF makes apparel available for purchase with slogans like "When's naptime?," "Sleep is Good," and "Sleepyhead!" A coffee mug with "Snooze or Lose" pairs well with an apron that reads "Gourmet Sleeper." These all mark a significant shift in language politics from a tee shirt that reads "Don't Sleep on the Importance of Sleep"—a rather staid slogan. Since its inception as an organization, the NSF has shifted its tactics from broadly sociological, conducting its Sleep in America polls to evidence the pervasive sleep disorders in the United States, to decidedly everyday politics.

Each of the weekly NSF alerts contains three or four announcements, primarily intended for a general audience. They invite readers in with their headlines, by asking, "Do Sleepy Toddlers Turn into Drug-Using Teens?," or by suggesting, "Sleep May Be a Predictor of Adolescents' Self-Esteem." The assumed audience of this information includes parents, teachers, physicians, and other authority figures in the lives of children. How could any responsible adult not take sleep seriously when the possible side effects include poor grades, depression, low self-esteem, and adolescent substance abuse? The alerts are precisely what they claim to be: alarming information for use in the everyday lives of individuals. The data have been extracted from their context to such a degree that they might be applied in virtually any context. Given the broad set of symptoms that lack of sleep is associated with, it can also be interpreted in reverse fashion, from symptom to cause. The uses of language such as "less sleep over time" and "sleep problems" combine to provide a vague basis for concern. How does one account for a window as broad as "over time" or what constitutes the beginnings of sleep problems? This abstract quality of the alerts is the basis of both their power and their frustration: at once, sleep's effects can be seen, but what is one to do about rearranging school-aged children's sleeping schedule? By late in the first decade of this new century, the NSF had adopted an RSS (Rich Site Summary) feed format for the alerts and began to send out updates whenever information was newsworthy. Along with the occasional sociological report, the foundation now focuses its attention on lifestyle concerns, asking, "Is there a perfect time to take a nap?" and suggesting that "cherry juice could affect your insomnia." Whereas the alerts formerly used broad populations and identified vague causal forces for impending social problems to appeal to those already concerned with sleep, now they focus largely on appealing to an audience who seeks their authoritative knowledge for the management of a sleep-friendly lifestyle. Adult napping strategies and nonpharmaceutical approaches to insomnia are a far cry from adolescent substance abuse, low self-esteem, and depression.

Time to Sleep, with P.J. Bear utilizes a similar tactic, although children are its target. Despite the anthropomorphic animals involved, the comic is decidedly educational: its tone is didactic, and the main characters spend too much time in exposition. P.J. Bear, with the help of his friend Rudy the Rooster, demonstrates how playing video games, drinking soda, and eating candy before bed might disrupt sleep. Rudy serves as a negative example, sleeping through his duties on the farm, leading P.J. to list the positive effects of a full night's sleep. They include more energy for sports and playing, a greater capacity for learning and memory, an improved immune system, more self-confidence, and heightened focus and attention; in addition, the "eyes, face and skin look healthier." The children portrayed in the comic, who are the recipients of P.J.'s advice, are ethnically diverse and evenly male and female, allowing, presumably, a wide array of children to identify with the "normal" children in the comic and to strive to incorporate P.J.'s advice into their lives.

Rudy the Rooter is the wayward sleeper who serves as an example for the ill effects of bad sleep hygiene. Rudy plies himself with great quantities of sweets; cake, candy bars, muffins, and soda are all apparent on the tray of food he's prepared for himself. This serves as his fuel for a night of video game playing. When chided by P.J. to go to bed, Rudy insists that after "one more game" he will finally be ready. But that last game never comes. What could be a more extreme negative example than a rooster who fails to wake his farm? This is Rudy's fate after a night of binge eating and video games. While some readers of *Time to Sleep, with P.J. Bear* might identify with aspects of Rudy's behavior (staying up late to play video games or watch TV, eating too many sugary treats), they are ultimately precluded from full identification with Rudy specifically because he is a rooster rather than a human. Rudy is a perfect abstraction: as a nonhuman representative of human behaviors, he serves as an aggregate of data—in this case, bad sleep hygiene and its consequences. Whereas the NSF alerts expect adults to see the abstractions represented to them in the lives of those around them and in their own sleep, *Time to Sleep* depends on children to see themselves in Rudy. Since they can only partly identify with him, since they can be roosters only in the most figurative sense, Rudy provides an abstract model that any child can identify with and no child can be subsumed in.

For abstract data to be real, for them to materialize and have consequences, they must be reembodied in the lives of individuals. This requires readers—the audience of the NSF alerts and *Time to Sleep*—to see themselves or those they care for in the abstract information being presented. Both of these media depend upon vagueness, but the means used to achieve this vary: in the alerts, vagueness depends on the translation of data based on small populations into generally applicable knowledge; in the case of *Time to Sleep*, it depends on the portrayal of specific behaviors through a character who is not too overdetermined for every child to see himself or herself in Rudy's lifestyle. The presence of a multiethnic cast of children, all of whom implicitly accept P.J.'s wisdom, provides a means for children to see themselves as part of the population who are at risk of developing Rudy-like behaviors. Moreover, the children are drawn in such a way as to be virtually featureless beyond their ethnic markers (particularly skin and hair color and eye shape), allowing for broad possibilities of identification. This process of identification through normative ideals is vital to widening the network in which sleep resonates: only the data needed for sleep to adhere should be mobilized; anything more than that might limit the possibilities for circulating information about sleep. Examples that are too particular, too specific, narrow the interpretive practices of the media's consumers. Conversely, abstract examples allow for both interpretive possibilities and the presence of doubt.

The process of abstraction and reembodiment relies upon the presence of normative ideals. The rise of statistical methods and census technologies has laid the groundwork for the constitution of norms. Trends and averages can be identified by collecting and analyzing a broad set of data, abstracting it through the use of qualitative markers like ethnicity, age, gender, and economic class.[2] Along with this has come the solidification of norms, the ranges of possibility for individuals identified by these qualitative markers.[3] And while the use of norms is primarily diagnostic ("this is how adolescents sleep"), the presence of norms can also become proscriptive ("this is how adolescents *should* sleep"). This is decidedly different from cultural ideals, which serve as concrete models to measure variance from them. If Thomas Edison is the ideal American sleeper—short nights in bed, quick naps throughout the day, both contributing to relentless invention and business acumen—he provides a model that no other American can achieve. Instead, one's inability to achieve the standard set by Edison provides the basis to understand the *cannot* of ideals. Whereas norms become imperative, ideals are aspirational yet prohibitive: no one can achieve them. Rather than conceiving of these modes of thinking as distinct epochal idioms (that ideals were premodern and norms modern), contemporary American medicine demonstrates how they are inextricably linked in the form of normative ideals. Norms have come to be understood as ideals: the average of eight hours of nightly sleep structures the pursuits of both individual sleepers and clinicians. Normative ideals are cultural models that structure ideas about being normal. Rather than being imperative or prohibitive, normative ideals are aspirational: they provide an ever-receding horizon of normative possibility, tangible because it appears to be normal. At once abstract and specific, normative ideals bridge the lives of individuals and cultural expectations of how everyday life can be lived.

Applying Abstractions

Normative ideals provide the basis for therapeutic normalcy. The aspirational components of normative ideals propel individuals on trajectories that seek interventions that ideally result in a normative state. This provides the basis for medical treatment, which takes therapy as its model: regular, periodic pharmaceutical, prosthetic, or behavioral adjustments in the lives of individuals. This model differs from curative models, which seek to affect a change in the symptoms of an individual that are lasting and require no further treatment. Therapeutic normalcy, in contrast, demands ongoing interventions, which necessitate patients' and physicians' assessment of the effectiveness of treatments and the alleviation of symptoms. Disorders can only be assuaged, not cured. Any treatment has the potential to lose its effectiveness, thereby necessitating further assessment and intervention. Normative ideals of spatiotemporal order—work, school, and family time—provide steady frameworks in which to attempt to structure disorder. This can be seen in Chip's case, discussed in chapter 1, and in many of the problems identified and interventions necessitated by clinicians in their assessment of patients' lives. In the following, to elaborate both normative ideals and therapeutic normalcy, I turn to the cases of patients as discussed by physicians at the MSDC.

It was not uncommon for parents to bring even very young children into the clinic for consultation. In one case, a three-year-old boy was brought to one of the staff pediatricians because the child was having frequent night anxieties. The boy would lie awake in bed after being put there, which would eventually result in his getting out of bed and seeking consolation from his parents. The attending pediatrician explained to the parents that they were putting their son to bed too early, that he would lie awake in bed, and lying there in the dark led him to begin to have anxieties common to children, those involving creaky noises, monsters, and fear of the dark. In order to alleviate these anxieties, the parents needed to change their "inappropriately early bedtime" for the boy, adjusting it to when the boy actually felt sleepy, not when the parents wanted him to be in bed. Further, the pediatrician suggested that the parents sleep in his room with him for the first part of the night to ensure that he get to sleep with minimal anxiety.

Immediately after presenting this case, the same pediatrician discussed the case of a seven-year-old girl who exhibited nocturnal arousals that were marked by a confused state and often screaming. When the girl woke up screaming, her brother, with whom she shared a room, would go and tell their parents. After some clinical detective work, these confused, screaming arousals were discovered to occur when the child was sleep deprived. The attending pediatrician explained that the girl's parents were part of a devout Christian group that required them to go to church at night twice a week. On those nights, the children would get less than their desired amount of sleep, because they needed to be at school at a set time the next day. The nights following those of church attendance were when the girl had her screaming arousals. Since leaving the child home during church service was not an option, nor was altering the girl's school commitments, the attending pediatrician planned to teach the girl relaxation techniques, hoping that this would be enough to help her settle herself before sleep and that this would curb the attacks.

Both of these situations involve the entire family, but they focus on the children. The children are seen as the source of the disorder, not the family as a whole or the children's other institutional commitments. This can also be seen in the case of a family of "owls"—late to bed, late to rise. The whole family reported regularly going to bed around 4 A.M. and waking around noon. The twenty-something daughter of the two late-to-bed parents recently gave birth to twins. The twins, by their mother's account, are both "larks"—early to bed and early to rise. The mother came seeking treatment for her own late sleeping. What was immediately available to her was a treatment similar to Ted's, discussed above, including set early bedtimes and exposure to light upon awakening. One of the pediatricians suggested that the twins would be the cure for their mother's late sleeping: their demands would force her to adjust to their schedule. To this, other physicians suggested limiting light exposure for the mother in the evening, which might also help her feel sleepy earlier.

In each of these cases, normative ideals structure the desires for sleep of both children and adults. When a family should sleep is shaped as much by the choices of individuals as the demands of institutions. In order to meet these demands, individuals and families subject themselves to rhythmic interventions in the form of prosthetic, pharmaceutical, and behavioral therapies. These therapies offer no terminal curative power: the symptoms will return if the therapies are ignored. Medical professionals' concern with patient compliance with treatment indexes this intensified regime of therapeutic normalcy. But compliance is far beyond simply taking a medication: because these therapies depend upon persistent and periodic use to be effective, they enroll individuals, families, and social networks in normative everyday formations. At the center of these social formations is the expectation that normalcy will result and that medicine will promote that result. This necessitates reconceptualizing what medicine is and does. Allopathic medicine is no longer about difference in the strict sense: increasingly, in accordance with the presumptions of normative ideals and therapeutic normalcy, allopathic medicine has shifted toward an interest in sameness, in integralism, in individuated control of the masses.

From Allopathic to Integral Medicine

Contemporary everyday life is inextricable from medicine; allopathy has enabled modern social forms through sanitation, inoculation, and pharmaceuticals that minimize exposure to risks and help alleviate disorders when we become afflicted. These powers of medicine are not self-evident: they insinuate themselves into society through the careful work of physicians, public health campaigns, and pharmaceutical advertising. Every medical invention depends upon this insinuative work to establish its place in society; this can require the promotion of awareness around particular individual and social disorders. It can also depend upon the invention of new disorders, novel combinations of symptoms or recodings of extant symptoms into isolatable disorders. This is the case with "excessive daytime sleepiness," which was once a symptom of narcolepsy and is now a separate disorder, treatable with stimulants. On a much broader scale, in the late 1990s and early years of the 2000s, the direct-to-consumer marketing around insomnia (centered on Lunesta, Ambien, and Rozerem) renewed Americans' interest in this disorder and offered novel solutions. The media of the National Sleep Foundation, the advertising campaigns for the Z-drugs, Requip, and Provigil, national health campaigns to reduce obesity, Internet support groups like Talk about Sleep, and clinical practice are all part of the same insinuative efforts. Because our everyday lives are saturated with media, these efforts at insinuation lie beneath our awareness, much like the powers and necessity of medicine. The successful insinuation of medicine into American everyday life has resulted in our generally not noticing or remarking on the powers of medicine to remedy disorders. But this insinuation is not without its causes, methods, and ends.

Insinuation depends upon the abstraction of data from individual cases. From these individuals, populations are produced: masses of insomniacs, narcoleptics, those with restless legs, and "excessive daytime" sleepers. With the isolation of abstract data in the form of a norm based upon statistical and probabilistic knowledge (the disorder's genesis, its usual course, viable treatments, comorbidities), these abstract data are reapplied to individuals. This occurs in two forms: the clinical encounter, in which a physician applies abstract data to an individual; and, secondarily, an individual applying abstractions to him- or herself. In either case, the individual is then rendered as an iteration of the masses, as one of many. The purchase of these insinuative powers has only increased over time, and with the contemporary direct-to-consumer marketing in the United States, as well as the proliferation of health-related Web sites and support groups, the insinuation of medicine into everyday life has become a commonplace in contemporary American society. But, more important, the insinuation of medicine's powers into our everyday lives forms society; it lays the basis for social interactions. This is evident in both the NSF's *Time to Sleep* and its weekly alerts: the NSF is attempting to make an American society in which it has influence and authority. To do so, it relies on both the production of data that are broadly applicable (about the sleep of others and our own lifestyles) and the presentation of data in a medium that allows readers to make it concrete, to see themselves as disordered sleepers. Mediation renders the abstract material, whether through individual consumption or expert diagnosis. And, at their base, these efforts of insinuation are outgrowths of integralism in medicine: attempts to produce slumbering masses and the data and formations to sustain them.

In this process of forming the masses, integralism affects three domains of life: the division of the social and the natural, the spatio-temporal rhythms of everyday life, and the pathologization of variation. How this is achieved is through two modes of insinuation: the integration of nonallopathic healing traditions into biomedicine and the intensification of everyday life as necessarily medical. Doubt, not market forces, is the basis of integralism; rather, capitalism is an effect of integralism, not its precondition. Integralism tends toward hegemony, affected through these alterations in the domains of life. But rather than a stable hegemony based upon established certainties and power relations, integralism's doubt proliferates emergent forms of life, leading to ever-new norms. This forms, in turn, new desires aligned to the everyday formations of integralism. These desires play at the limits of our human capacities, making and remaking the qualities of being human. What we accept as medicine, as American allopathic medicine, is a much more complex beast than we often allow and is indebted in its motives to its central doubts—about the human, about its powers, and about its ends.

Allopathic medicine is particularly an *American* medicine; the histories of this medicine and of the United States are intimately intertwined, with the institution of allopathic medicine and the nation developing together from the eighteenth century through the present. Allopathic medical practice came of age at the end of the nineteenth century through its competition with homeopathic medicine, differentiating itself through a fundamental assumption of treating illnesses with substances unlike the illness being treated. Homeopathy, instead, has based its practice on qualities of likeness. Its basic conception of effectiveness is to apply a small dose of the same thing that causes a condition in order to effect the opposite condition. For example, since caffeine promotes wakefulness, one way to remedy insomnia is with highly diluted caffeine. This varies from allopathic practice, which treats insomnia with something that produces sleepiness, not wakefulness. Becoming increasingly aligned with scientific laboratory practice in the beginning of the twentieth century, allopathic medicine secured its role in both university education and the care of the ill. With the rise of industrial chemical production, the development of national and international trade networks, the establishment of the Food and Drug Administration, and the commercialization of pharmaceuticals, allopathic medicine became the standard for American medical practice. Along the way, it came to be known as "biomedicine," referring to its basis in laboratory biology, and then, increasingly, known simply as "medicine." This has occurred alongside the constant presence of "alternative," "complementary," and "folk" medicines, which, for many Americans, continue to provide primary or supplemental treatments for everyday medical complaints.[4] Moreover, these alternatives often provide a sense of hope when allopathic treatments prove ineffective or noxious and unable to produce the desired outcomes. What characterizes allopathic medicine from other healing traditions is its reliance upon difference, which comes in two forms: how individuals in their disordered state are different from themselves in their prior, healthy state and how individuals are different from populations. In both cases, allopathic medicine relies on control strategies to return individuals to normal—normalizing them either in relation to their own baseline health or against population statistics. Individuals come to rely on allopathic medicine for solving their differences through the insinuation of its authority through the media and institutions that form our everyday lives.

Early in the twenty-first century, allopathic medicine is addressing the threat and promise that the presence of alternative traditions offers, paralleling American concerns with multiculturalism, immigration, and postcolonial transnationalism. Alternative medicine has latterly returned as a component of allopathic medicine, sometimes referred to as "complementary," "holistic," or "integrated" medicine, including practices such as acupuncture and chiropractic as well as "wellness" and "mindfulness" programs. Integrated, complementary, and alternative medicines assume two or more distinct bodies of knowledge and practice—allopathy and its complementary practices—but the history and contemporary ethnography of medicine show how muddled the ontological, epistemological, and practical bases of all medicines are: allopathic medicine has had a profound effect on other medical traditions and vice versa.[5] The drive toward inclusion of these complementary forms marks allopathic medicine as *integral* medicine, a medical paradigm and practice that has motives other than simply the curing of ills through the deployment of difference; instead, integral medicine is about producing the masses, populations subject to control through individuated means. One way that this is accomplished is through an ever-proliferating set of techniques, recognized as medicine, with health as their object, and complementary to contemporary capitalist formations. What makes allopathic medicine integral medicine is not its relationship with other medical traditions but its basis in spatiotemporal rhythms and normalcy that is predicated upon assumptions of nature and society and that takes as their end the formation of masses subject to medical control. This process of forming the masses is hegemonic; integral medicine is totalizing, bringing individuals, techniques, and treatments into a unified everyday formation that defines its capacities through reliance upon the normal. The masses, the populations, produced by integral medicine are also always markets— laboring bodies that are also consumers—and they interact with medicine through capitalist exchange, including medical insurance, medical care by professionals, and the purchase and use of pharmaceuticals.

At the turn of the twenty-first century, American everyday life is fundamentally tied to ideas of the rhythmic, emblematized in the repetitive use of pharmaceuticals, which produce and depend upon everyday notions of time and space. When a rhythm is so vital, its disordering can have profound effects. This rhythm is based on normative expectations about sleeping and everyday life and, as such, might simply be taken as an intensification of the everyday rhythms that many Americans abide by to structure their social and biological desires. The expectations of any of these arrangements are necessarily that human biology be brought into accord with the spatiotemporal formations of American society, with the cultural expectations of American sleepers regarding normal sleep, and with the veracity of pharmaceuticals as a curative for their irregular, disordered rhythms. This is why, rather than refer to aberrations of sleep as illness, disease, or disability, I use the term *disorder;* integral medicine expects and takes as its object the production of order. When bodies resist this ordering, their disorderly conduct or being is what legitimates the intervention of integral medicine.

Integral medicine works through the establishment of natural hegemonies, of rendering individuals and masses isomorphic in their qualities. This is often accomplished by naturalizing institutional formations, as in the conception of the workday as founded in primordial human relationships with time and space. This process has been attended to by a number of historians and social scientists, who have looked at the naturalization of social norms and cultural expectations, often referred to in the Marxist tradition as reification.[6] Scholars have identified reification as the process in which inequalities and assumptions about race and ethnicity, class, gender and sexuality, age, and susceptibility become concretized, leading to the further compounding of inequalities and assumptions about individuals and groups.[7] This enfolding of cultural assumptions into nature leads to the centrality of uncertainty in integral medicine, namely, doubts about what appears to be social and what appears to be natural. In this slippage resides one of the motives of integral medicine: the need to convey the resolute nature of what might otherwise appear social, thereby propelling the need for hegemonic efforts on the part of individuals and institutions as they work to establish the social as natural. By rendering everything natural instead of allowing for the possibility that the social influences or determines health, illness, and behavior, we make it possible for medicine and science to act upon the pathological and for individuals to represent the masses. To allow for the possibility that society has determined or generated the illness or bad behaviors of individuals is to render action out of bounds for medicine and science and to put it into the hands of governments, the law, and other institutions, which take as their object the control and ordering of society through the management of behaviors. One might think here of the recent controversies around mental illness as produced by aberrant brain chemistries or malformations of the brain. Rather than accept that mental illness might be socially produced, leading to brain chemistry imbalances or the retardation of neurological development, we emphasize genetic causes as leading to mental illness. This "biology as ideology" provides medical practitioners with authority over mental illness, its treatments, and what counts as legitimate claims to disease.[8] These assumptions justify the use of medicine as a form of contemporary control of the natural and draw on American medicine's colonial and industrial legacies of surveillance and the control of individuals and masses.

This control is the basis of our desires and intimacies. How we conceive of ourselves—as sleepers as well as individuals—is through our relationships with other individuals and the institutions that make up our everyday lives. These are my concerns in Part II, which explores the many processes of integralism in medicine through individuals and their interactions with everyday institutions. Through their interactions, they come to know their sleep and wakefulness; they come to know how to manage themselves as somnolent subjects. In their efforts and in the efforts of institutional actors, emergent intimacies are founded: among individuals, between individuals and their therapies, and between individuals and spatiotemporal rhythms. Recall Ted, who opened this chapter. For Ted, his control of sleep and wakefulness was tied to his desire for the same, his desire to learn. That he came to find himself pedaling a stationary bicycle through his mornings at school created a new intimacy: with himself as a sleeper, with the bicycle, with his school, peers, and teacher, with his parents. These new conditions of his life became an inextricable web of intimacies and desires that could be interrupted only with substitutions in order to maintain the order that had been established. Although we may feel far from Ted, we are all like Ted, and the intimacies that constitute our everyday lives sustain and fatigue us, motivate and burden us. And, fundamentally, they tie us to the everyday and the unfolding demands of integralism and its doubts.

The power of medicine lies in its ability to cure, in its ability to make bodies anew; the force of medicine is homogenizing: it makes bodies the same, it produces the masses. Consider the case of Dee, a forty-six-year-old black woman who had sought clinical treatment for her chronic insomnia about a year previous to our interview. Dee was self-employed: she owned a small hair and nail salon with a steady clientele. For the previous three years, Dee had experienced increasingly disrupted sleep and ever-later times of sleep onset. By the time we talked, she routinely went to bed around 4 A.M., which made opening her salon by 9 A.M. difficult. She started missing clients because of their early schedules and her late one. She purposely scheduled long lunches for herself and slept in her office on a thin mattress. Her workdays ended up extending rather late with paperwork, paying bills, and managing deliveries. "I wasn't anxious at first," she explained, "but when the business started going bad, my sleep got even worse." Dee went on:

> At first they thought it was just insomnia. Then they decided that it was premenopausal. That was after they had given me a prescription for Ambien, and it didn't really help anything. It actually made things worse, because I still couldn't sleep when I took it, and then I would just feel groggy and hazy all night, until I fell asleep.

Dee's children were both grown and in college; her husband worked evenings at a local manufacturing plant. After being unable to tolerate the effects of Ambien, Dee eventually stopped taking her medication; the fix, she decided, was to change the hours of the salon. She hired someone to handle the few clients who were unable to adjust their schedules to Dee's, and she serviced the rest of her clientele in the afternoon and evening.

The intention for Dee's medication while she took it was to move her to a normalized schedule, in terms of her work, sleep, and family life. Her doctors worked to align Dee with the spatiotemporal norms of American everyday life, despite the possibility of alternative social orders that would not have pathologized her behavior. Medicine attempted to integrate Dee with society. It promised her a new spatiotemporal rhythm, a new understanding of the social and natural causes of her sleep, and measured her against population norms. Dee could choose not to medicate her disorder—as she did—but this entailed her being out of synch with dominant American spatiotemporal rhythms. For some, this is bearable; for others, medicine provides the means to reenter society, to be orderly once again. For Dee, the therapy was less tolerable than being out of synch with society.

As medical thinking is increasingly disseminated through everyday forms of media—not specialized journals, but Web sites, comic books, pamphlets, and advertisements—Americans are invited to think of themselves *through* medicine. Did medicine offer Dee a solution? Yes, but it was not the only solution, and, for Dee, it was not the most appropriate solution. But she turned to medicine nonetheless because of the explanatory power that medicine has in American everyday life. In the cases of media presented herein, sleep is an explicit concern, and these media might be targeting those who already think of themselves as troubled sleepers, but many messages about medicine are more discreet: they pervade our daily lives and go largely unnoticed. When it comes time to making sense of our disorderly experiences, however, medicine offers one means to do so. In seeing ourselves through medicine, we integrate ourselves into the body of the masses—that abstract set of data that rules normative expectations about bodies and their behaviors. In so doing, we eradicate our differences by becoming subject to the power of medicine. It is only when bodies react poorly to treatments, when individuals become noncompliant, that difference is reasserted. However, increasingly this difference is pathological too: noncompliance is taken to be dangerous to the individual prescribed with the treatment, as well as to society more generally. As medicine comes to govern and constitute lives, our ability to think outside medicine, to conceptualize ourselves beyond the masses, becomes an obscured luxury. Integral medicine, in its very function, dominates life itself and determines our individual desires as much as it lays the basis for society. In its hegemony, it both obfuscates its basis in doubt and depends on ever-proliferating questions of life and desire.

II.

CULTURES OF SLEEP

4

....

Desiring a Good Night's Sleep
Order and Disorder in Everyday Life

IN THIS CHAPTER, I CHANGE MY FOCUS TO EXAMINE THE LIVES
of individuals and families as they are affected by disorderly sleep. I
move from the disciplinary and institutional logics of sleep science and
medicine and attend the forms of life that they produce, largely in the
lives of individuals who see themselves in need of medical intervention.
I am interested in the lives that individuals form for themselves and
the effect they have on those around them. In order to establish their
everyday lives as orderly, individuals integrate new pharmaceuticals,
prosthetics, and behaviors into their self-care. These intimacies, these
alliances with new objects, form emergent desires.[1] In some cases, these
therapeutic intimacies are intolerable for individuals, who struggle to
shape their everyday lives through varying means, sometimes opting out
of medical treatment altogether. But there are similarities across individuals diagnosed with the same disorder, both in treatments available
to them and in the consequences of finding nonmedical alternatives to
managing their everyday lives. In relation to these similarities, I structure this chapter nosologically, by medical definitions of disorder. In
so doing, I bring together cases of individuals and families who share
symptoms but who resolve their disorders differently from one another.
In some cases, their choices involve medical treatment. In other cases,
individuals find nonmedical means to allay their sleep disorders. I draw
on cases similar to these throughout the rest of Part II, examining how

individuals with similar sleep disorders and their families form and are formed by the institutions they encounter, in particularly American ways, through their emergent relations with therapies, institutions, and the desire for sleep.

Early twenty-first-century American society has been marked with an intensified interest in sleep as an explanatory device for health and its impediments, a trend that differs substantially from medical discourses of the 1990s, in which sleep was largely ignored. This change has been brought about, in no small part, through direct-to-consumer advertising of sleep-inducing medications, now annually a multibillion dollar industry, as well as through the efforts of various individuals and institutions to popularize sleep. My interest in this chapter, in tracking this shift during the 1990s and the early 2000s, is to attend primarily to the alterations in treatment regimes, namely, the intensification of pharmaceutical therapies for health problems, and to elaborate what might be particular to this seemingly emergent pharmaceutical aspect of integral medicine's further development. "Pharmaceutical personhood" has become an ever more naturalized form of life in contemporary American society and in modern life more generally.[2] Beyond the reliance upon medications as a means of alleviation—which is new only in the sense of its intensification, inasmuch as medications and medicinal supplements have a protracted history—my interests are specifically in the "time–discipline" of pharmaceuticals,[3] the ways that the use of pharmaceuticals produces particular kinds of spatiotemporal predispositions, desires, and intimacies and how these form and are formed by American everyday life. I see in pharmaceuticals and pharmaceutical use a reliance upon repetition, the formation of everyday rhythms that reinstate themselves through desirous means. Simultaneously, pharmaceutical regimes are also constituted by ideas about the everyday, about rhythm and spatiotemporal order. Contemporary pharmaceutical use depends upon a mode of enframing that produces the spatiotemporality of the everyday inasmuch as it mitigates disorder through the formation of normative orders of desire and intimacy. In the early twenty-first century, American everyday life is tied to ideas of the therapeutic, emblematized in the repetitive use of pharmaceuticals. When a particular everyday rhythm is so vital, its disordering can have profound effects for individuals, their families, and society more generally, and order can be reasserted by turning to the powers of medicine.

"We Don't Get Enough Sleep"

Marcus and Laura Burton were professional, upper-middle-class, white Americans from the suburbs of Minneapolis and, at the time of their interview, were in their early fifties. They had been together since their early twenties, having met as undergraduates while at a university in the Midwest. They were not unique in their sleep problems; in fact, neither was recognized as having a nosologically defined sleep disorder in the strict sense, other than being "tired, basically." As they explained, they came to identify their health problems *primarily* as sleep problems, but the regulation of sleep was a difficult project for them, and social obligations often took precedence over desire for sleep. Marcus worked as a senior engineer in a medium-sized firm that he was part owner of, and Laura worked in a state-run hospital in various community outreach programs. Their two daughters were fully grown, graduated from college, and in the process of establishing their careers and families. Living in their empty nest, the Burtons had begun to tend to their sleep complaints, which now stood in stark relief to their other social obligations. What the Burtons evidence, and what they were clearly aware of, is that sleep disorders are also inevitably social disorders and that their health complaints affected both themselves as individuals and each other as bed partners. Their desire for sleep interfered with interpersonal intimacy, and this was related to Marcus's close relationship with stimulants. What marked the Burtons' interaction with their sleeping problems was a generous humor, directed both at each other as sleepers and spouses and at the role that sleep played, or failed to play, in their daily lives. This use of humor both deflected the force of sleep's disruptions from their daily lives, rendering it a sort of joking nuisance rather than an unrelenting irritation, and acted as a point through which their marriage was worked and reworked: desire for sleep became interpreted as another necessary aspect of their relationship as spouses and as bed partners, bringing together broad social obligations, cultural expectations of space, time, marriage, consumption practices, and their intersubjective biologies in an intimate chemistry of desire.

What were you diagnosed with?

MARCUS BURTON: Well, it turns out I was basically diagnosed as tired. And—if I remember the specifics of it—he said on average I had 2.2 apnea events per hour, which didn't qualify as anything dramatic by any stretch of the imagination.

LAURA BURTON: It was actually under normal.

MARCUS BURTON: Yeah, what he said was less than normal. And my sleep was 87 or 88 percent efficient, as opposed to 82 percent for the population in general. And my favorite part of the whole thing was the nap assignment. And I think it was because I didn't have any significant apnea overnight they wanted to see what my sleep pattern during the day would be. So, when I finally arose at a quarter to ten, they gave me the prescription—if you will—to take five naps during the day, and I did that at 12, 2, 4, 6, and 7 [P.M.] (they decided they didn't need to wait until 8 for the last one, as I can fall asleep pretty easily). And as the day went on, I think I fell—I think it took a little longer to fall asleep each time. I think it took—

LAURA BURTON: But he fell asleep each time.

MARCUS BURTON: Yeah—doctor's orders. I think the last time it was twelve minutes it took me to fall asleep; the first time I think it was two and a half. But I was pretty tired. And the long and the short of it was I was tired, basically. I don't get enough sleep—we know that: We don't get enough sleep.

The Burtons, by all accounts, are "normal." They are representative of a broader phenomenon in the contemporary United States, namely, that of a public in the making. The contours of this public are biological, and they work through and rework ideas about human biology and its relationships with society and culture on the basis of the formation of new desires and intimacies. A necessary component of this emergent biological public is the simultaneous elaboration of therapeutic normalcy. Studies of normalcy primarily focus on the means by which norms are enforced through institutional expectations and discipline and on how individuals interpellate themselves into normative forms of desire, into the masses.[4] What is most often privileged in analyses of the normal in medicine are the end-focused effects of norm attainment; that is, once individuals become normal or are resigned to their abnormality, the analysis ends. However, integral to the contemporary pharmaceutical society Americans find themselves to be mired in, therapeutic normalcy depends on a practice of medicine that can never reach its end. Therapeutic normalcy depends on an open, never-ending, unachievable normal that individuals aspire to—in this case, the normative ideal of eight consolidated hours of peaceful, natural sleep. Rather than leading to a therapeutic end, which might be understood as a "cure" (a temporally specific, end-focused process), pharmaceuticals depend on repetitive acts: days punctuated with medication, monthly refills of prescriptions, and annual checkups, resulting in both self-surveillance and institutionally based observation. In so doing, the clinic, rather than being a stable space, has become an ambient background of American everyday life, and one's disorders of sleep are understood in the idiom of "health." The Burtons help explicate these ideas further as they apply to "normal" Americans.

Marcus's perceived abnormal sleep—high efficiency, few apnea events—was not a problem for the Burtons; rather, the problem was what had been increasingly taken as his normal sleep, or lack thereof. Marcus exhibited chronic sleep deprivation, which led first to his ability to test as "tired" and to nap on cue and second to his and Laura's recognition of the deprivation as foundational to his continued health. Marcus and Laura were aware of the problems that sleep had caused in their lives and in their marriage, and this awareness is what motivated them to seek medical treatment, although the possibility that they could arrange their social lives in ways more amenable to sleeping was also available to them, however undesirable. Despite knowing that medicine might have solved their sleep problems, Marcus had deferred medical help, in part because his life had felt too "complicated." What brought him back to the possibility of a medical intervention was preserving his marriage, as well as an ongoing understanding of his body's rhythms as disruptive for himself and Laura.

What led you to seek medical treatment?

MARCUS BURTON: Well, Laura, and that I know that I snore. And I have had years of listening to my father and my brothers [snore], you know, as a child. And I know how loud it is. And I also know what it's like to be sleep deprived in that I went through a phase of my life where I had a real hard time sleeping for about a month, and life's no fun when you're not sleeping. And, so, when Laura said, "It's really hard to sleep if you're making noise like that," and one day she got up and left—

LAURA BURTON: I went into the room across the hall.

MARCUS BURTON: You went across the hall and set up a camping cot and went to sleep. . . . I don't remember the exact chronology, but I had signed up for my sleep study the first time two or three years ago, so it wasn't something that I had—that I fought or didn't buy into, but I had gone to my GP, and he told me to contact some sleep institute. Laura was working at MCMC, and she said, "You can't go there; you have to go to MCMC," so I canceled the other one, and life got complicated, and two years went by. . . . She's in menopause, she has insomnia to begin with, and—

LAURA BURTON: I think that was what really brought you there, because I mean we've been married and sleeping together for thirty odd years, and I've been able to sleep through it until recently. And because my sleep cycle changed, my quality of sleep changed; unless I'm really, really tired, I can't fall asleep with him snarking in my ear. I just can't. But I think that's basically what happened. I think he's been constant, and it's me that changed.

Worth noting is that it was not so much a change in Marcus's condition that finally brought him to seek medical help but rather a change in Laura's body, her ability to sleep and tolerate Marcus's otherwise constant snoring (or "snarking" in their idiom) body. Only when Marcus's sleep disorder impinged upon their intimacy and desire for sleep did he finally seek a medical intervention. This disorder of intimacy, in turn, was based upon the interaction of Marcus's and Laura's desires for sleep; when the balance that had previously existed became too disrupted, therapy was necessitated.

Caffeine became a means of self-medication for Marcus, working in turn to conceal his sleep deprivation and, as the Burtons recount, also exacerbating the disorder generated by his "snarking."

LAURA BURTON: Did you talk about the coffee?

MARCUS BURTON: No. And, um, what she means by that is when I, when we, had the initial interview with the doctor he said, "Do you drink coffee?" and I said, "Yes," and he said, *"Well, how much?"* and

I said, "Two, three pots a day." He said, "You mean cups?" "No, I mean pots, and, uh, but I always stop by about 3:30—I never drink coffee after 3:30." And that led him even further down the pipe to his conclusion that "Well, that you're really sleep deprived." He had a resident with him at the time—he was training somebody in—and said, "It's [caffeine is] a very potent stimulant, but it doesn't last very long, so if you're sleep deprived but you need to stay awake for twelve hours, you might drink three pots of coffee in a day," and [laughter] he said, "You know, you might really want to stop that." And, so I finally had a doctor tell me I was drinking too much coffee. No doctor had told me that before, because my blood pressure isn't high, and I never actually killed anybody or anything like that, although it can make me pretty edgy. So I stopped. Instead of drinking about [a cup of coffee]—I would typically have about five or six of those by eleven o'clock in the morning, and probably about three or four more in the afternoon, and coffee at lunch wherever I was. I hacked that down to about one or one and a half in the morning and one in the afternoon, and I just made a conscious decision to stop.

LAURA BURTON: And it made an incredible difference in how much he snores. Because he snores, way, way, way less now.

MARCUS BURTON: Or you're sleeping better.

LAURA BURTON: No, I'm telling you, I'm not sleeping better. Your snoring has gotten better. Or gotten—I don't know if you call it improved snoring if it's quieter. It's not as loud; it's not as often.

MARCUS BURTON: I still wake myself up. I still wake myself up on occasion with a snark here or there. But we still don't get enough sleep.

The use of so much caffeine (a quantity not out of step with that of other disordered sleepers) not only concealed Marcus's sleepiness (although how effectively it did so is questionable, since he was still aware of his sleepiness) but also exacerbated his predisposition to be a noisy sleeper. His desire for stimulation, his intimate relationship with caffeine, disrupted both his and Laura's sleep. Marcus may have been an extreme caffeine drinker before his recent reduction in consumption, but his modified levels of coffee drinking are in line with reports of average American caffeine consumption, which is estimated at two hundred milligrams daily for noncoffee drinkers and nearly twice that for those who drink coffee.[5]

Like many Americans, if the National Sleep Foundation's Sleep in America polls are accurate, Marcus averaged less than eight hours of sleep per night. But whereas many Americans have reported difficulties with sleep onset, sometimes conflated with insomnia, Marcus manipulated his sleep duration to meet his everyday obligations, particularly work.

Can you describe a typical night's sleep before the diagnosis?

MARCUS BURTON: With me? Typically we would go to bed between twelve and one, and I would go to sleep in two minutes—

LAURA BURTON: Twenty seconds max.

MARCUS BURTON: Yeah. I would usually have to get up once in the night to hit the lavatory. I would typically get up around seven. I try and get at least six hours of sleep a night, knowing my ideal is seven. In terms of waking myself up snoring, to me it doesn't seem like a common phenomenon. But I know that it happens— I just couldn't—obviously I don't keep track on a nightly basis, but I would expect that I probably wake myself up once or twice a night. I realize that it has happened and roll over, or try to do something. I usually sleep on my back, and after the first snark of which I'm aware, I roll over to my stomach, because I'm convinced that it's physiologically impossible to snore while sleeping on my stomach.

While there is no proper medical treatment for Marcus's snoring, he and Laura worked to find means of minimizing its disruption of their sleep and realized ultimately that it was not simply his snoring that impacted their sleep but also a broader network of everyday demands and expectations. These everyday demands led them to modify their desire for sleep by developing intimacies with stimulants, everyday routines, and expectations of themselves and each other.

Thinking about oneself through one's desire for sleep and normalcy inevitably invites thinking about what kind of sleeper one is, whether a lark or an owl, light or sound, orderly or disordered. To know oneself through one's desires and rhythms of sleep is also to know how the spatiotemporal orders of American everyday life serve as either nuisances of or foundations for one's life. Moreover, knowing oneself as a sleeper also invites understanding difference among kinds of sleepers, including such nonnormative patterns for Americans as napping and day sleeping, the former being short periods of supplemental sleep, and the latter being consolidated daytime sleep. In the following excerpt, Marcus and Laura ruminate on the kinds of sleepers they are and how this has impacted their everyday lives:

> LAURA BURTON: All I know is that I've always been a night person, and I get a lot done at night.
>
> MARCUS BURTON: And I'm just somebody who doesn't get much sleep. So I'm—whether I'm a night person, or a day person, or a morning person, it's hard to say, but—
>
> LAURA BURTON: But I can power sleep, and you can't really do that. I mean, I can sleep for twelve hours.
>
> MARCUS BURTON: Oh yeah, I can't. I have to be absolutely dead exhausted just beyond all get-out. If I'm tired I sleep for nine or ten hours, if I'm not I sleep for seven. And when I'm on a normal workday, it's five and a half to six, usually.
>
> LAURA BURTON: And napwise—
>
> MARCUS BURTON: I can take naps and you can't.
>
> LAURA BURTON: Well, I can take naps, but twenty minutes isn't necessarily refreshing for me.

How common Marcus's and Laura's experiences as sleepers are is debatable, as is their status in relying upon sleep and its disorders as a means of understanding their individual subjectivities and intersubjective relations. However, their intimate knowledge of their own sleep habits and patterns as well as that of each other is profound. While they may not have been common at the time of our conversation, with their awareness, they offer an ideal type of the American sleeper at the turn of the twenty-first century: someone who not only perceives sleep as central to health and the ordering of everyday life but also conceives of her or his own desire for sleep and its nested intimacies as a vital means of thinking about oneself, social possibilities, and therapy as a means of mitigating possible concerns. If the Burtons are normal American sleepers—sleep deprived, wrapped up in intimate relations with institutions, chemicals, relationships, and seeking medical help for being "sleepy"—narcoleptics offer a *model* of American sleep, with alertness and sleep always mediated by intimate relationships with pharmaceuticals.

Narcolepsy: Modeling American Sleep

A middle-class white American, Sam was in his early forties at the time of his interview and had become aware of his sleeping problems—narcolepsy with cataplexy (a sudden loss of muscle tone)—in his midteens, which his family doctor at the time understood as hypoglycemia. As Sam described it, he "had a very sketchy parenting situation as a youth, so it was not really pursued as a problem then and nobody worried about it too much." He went on to explain that throughout his time in elementary school, the school nurse had treated his cataplexy events with naps supplemented with candy. As he aged, his cataplexy abated, but it was replaced with "excessive daytime sleepiness," which plagued his high school years and life thereafter and eventually led him to seek medical diagnosis. In Sam's words, "My father noticed my sleepiness and directly accused me of smoking pot or using pills. . . . He did not understand, and there was absolutely no way he would believe that it was not pot or drugs that made me sleep in class and during the daytime. No one could be that sleepy, or so he thought." Sam went on to explain that when he was in his early thirties his narcolepsy began to express itself fully, affecting his "ability to work and play." He sought medical help and was first diagnosed with depression. After a few years of treatment without the cessation of his symptoms, he sought new doctors. In Sam's words, "The new doctors were just like the old doctors, and they started all over, barking up the depression tree. I went through all the steps again (shrinks, new meds, sleep hygiene, et cetera) before I decided that the doctors maybe couldn't help me with this, so if I was going to get better I was going to have to figure this out on my own."

Sam then quit his job to tend to his health full time. He went from doctor to doctor, from one specialist to another, until he was eventually diagnosed with narcolepsy—twenty years after his symptoms had begun. He was then prescribed "Dexadrine as well as a few other types of speed and knockout drugs," explaining further, "But they all seemed to make me really sick." At the time of our interview, Sam had decided to stop taking medication for his narcolepsy, which he explained as "choosing not to get on the drug grind that most narcolepsy patients are on." In justifying his choice, he said, "Most of the meds are not yet tested long term, they make me ill, and I do not know any old speed users. I may sleep a lot, but I am *me* when I am awake, and not under the influence of meds. . . . I may have to use them again one day to function. But for now the challenge is for me and my narcolepsy doctor to figure out how to maximize my 'good' time."

What Sam refers to as "the drug grind" is a crisis of desire and intimacy.[6] It is a crisis that is spatiotemporally specific (it affects Sam at a moment in his life) yet has unfolding implications. It constitutes an event through which Sam's life will change and be marked. Sam is caught between possibilities: a life of medication and normalcy, undercut by feelings of unease; a life without medication but with social limitations; and variations along the spectrum formed by these two choices. Sam emerges from this crisis with new desires, new intimacies: a new form of life measured against the expectations of spatiotemporal order embedded in American everyday life. Because of the variety of forces and possibilities that this crisis is composed of, Sam's decision will always entail some doubts: Would another medication be better tolerated? Would some other management of behavior allow him to work? Would another doctor be able to find a more effective cure? This crisis of desire and intimacy brings together the social, the biological, the economic, and the chemical not only to produce treatments for health complaints but also to produce new complaints; this crisis of desire and intimacy both depends on old expectations of normal bodies and behaviors and produces new normative ideals of health and spatiotemporal orders of the everyday.

Consider, for example, the differences between the foundation of sleep medicine—the consolidated eight hours of sleep—and the actual sleep patterns of disordered sleepers. For every disordered sleeper and his or her bed partner, what is of primary concern is a lack of sleep. Increasingly, medical insurance plans in the United States have recognized the need to cover sleep studies and pharmaceuticals to normalize sleep. And to ensure that patients are properly diagnosed, the American Academy of Sleep Medicine requires accreditation of sleep professionals and their clinics, working to insinuate their authority between medical professionals, medical insurance policies, and patients. Insurance plans cover some treatments and not others; clinicians provide some diagnoses and treatments and not others; and, most important, individuals respond to some treatments and not others. Between what is provided and what could be, on the one hand, and what is responded to positively by individuals, on the other, invariably doubts occur. And it is in this gap between the lives of individuals and the spatiotemporal formations of everyday life that most individuals find themselves caught between their sleep disorders and their social obligations, trapped between desires and intimacies.

Sam's choice should not be misconstrued as resistance to medicalization; it is, rather, an attempt to produce another order of desire and intimacy: there is nothing beyond desire, only other models of biological, social, cultural, economic, political, and moral ordering. The side effect of Sam's decision is that he is forced to live outside dominant society: at the time of his interview, he had never finished college and had spent the previous seven years unemployed because of his need for naps and his inability to be at work when required. He lived abiding by only his own spatiotemporal rhythm, and although this rhythm overlapped at times with the dominant spatiotemporal orders of American society, it failed to be so synchronized as to qualify him as a "normal" sleeper. If resistance can be read in Sam's refusal of medical treatment, it should be read as a rejection of the dominant form of life endorsed by contemporary capital interests that would require him to treat his disorderly sleep through pharmaceuticals, not a rejection of everyday life altogether.

In contradistinction to Sam is Kate, a white woman in her mid-fifties at the time of her interview. Her narcolepsy symptoms had begun at around age fifty-one, and she struggled through her local medical community over the ensuing two years, attempting to be properly diagnosed. In addition to the excessive daytime sleepiness often associated with narcolepsy, she also experienced hallucinations at sleep onset and frequent insomnia. In part this latter symptom had led to her first diagnosis of depression, but when the symptoms failed to resolve themselves after she was placed on antidepressants, she sought further help. After being diagnosed with narcolepsy, she was first placed on Provigil and Adderall, but she developed what appeared to her to be restless legs syndrome as a side effect of the drugs. In consultation with her doctor, she was placed on Xyrem, which resolved her narcolepsy symptoms and her RLS. Also, a desired side effect of the Xyrem, as she explained, was that it "structured" her daily life, because she needed to ensure that regardless of what she was doing she would be able to take her required medication. The difficulty she faced with her initial Xyrem prescription, however, was its dosage. The drug had originated from Jazz Pharmaceuticals, and her doctor, unfamiliar with it, depended on the authorities at Jazz to guide his dosage of Xyrem for Kate. Unfortunately, the company started her at a much higher dosage than her body could handle, and she quickly became uncomfortable with the drug's effects. This led her doctor to lower the dosage and gradually work his way back to near the expected dosage ceiling. Despite her problems with Xyrem, Kate referred to it as a "godsend."

The model narcoleptic subject is Martin, a twenty-one-year-old white man at the time of his interview. He had had narcolepsy symptoms since around his sixteenth birthday, including frequent cataplexy attacks. In addition to carrying prescriptions for both Xyrem and Provigil, Martin had also been diagnosed with bipolar disorder, which was medicated with Lamictal; with attention deficit disorder with hyperactivity, treated with Focalin XR; and with persistent panic attacks, treated with Paxil CR and Xanax XR. To improve his daytime alertness, Martin was also prescribed Adderall XR, which was often used for ADHD. He also reported that when needed, he would supplement this array of drugs with additional Focalin and Xanax tablets. In conversation, Martin estimated that the cost of his monthly prescriptions ran upward of three thousand dollars, a cost that was offset by his parents' medical insurance plan; as he discussed his future employment possibilities, Martin was certain that he needed to find an employer with a comprehensive health care plan. Amazingly, with his multiple prescriptions, he experienced no side effects, a feat that Martin ascribed to his primary sleep physician, a psychiatrist who specialized in sleep disorder patients who carried multiple diagnoses. Also rather exceptional was Martin's susceptibility to the many drugs: each of them seemed to have exactly the effect desired. Because of this efficacy, he was thoroughly ensconced in the spatiotemporal order that comprised the pharmaceutical, economic, and social obligations and commitments that maintained his treatments. The area where his narcolepsy seemed to be a continued source of tension was his love life, in which he was unable to keep a steady girlfriend, his most recent being unable to cope with his narcolepsy despite his pharmaceutical normalization.

What these three cases evidence is the varying degrees to which bodies tolerate medication, the degrees to which individuals are willing to resolve their sleep complaints by developing new desires and intimacies, and the degrees to which they are willing to accept dominant pharmaceutical answers to their experimentations with sleep and the spatiotemporal order of American everyday life. Sam and Martin tolerated the pharmaceutical treatments and their intimacies at opposite extremes of the spectrum. Neither resolution was entirely satisfactory: for Martin, the side effects of treatment circumscribed his dating life and his possible employment paths; for Sam, opting out of the pharmaceutical prescriptions for his narcolepsy entailed similar social impacts. Kate, however, seemed to be able to negotiate some middle ground in her pharmaceutical and intimate investments. Kate's negotiation of both the medical system and attempts to treat her symptoms eventually led to the correct diagnosis, but even then she needed to further negotiate the prescriptions given by her sleep doctor. With the right drug, Xyrem, Kate was able to return to her everyday social obligations, fully insinuating herself back into the dominant spatiotemporal order of everyday life. The drawback to Xyrem, as she noted, is the force the drug exerts on the daily life of its users, requiring them to take it at specific intervals, based on normative models of everyday life.

Unfortunately, for some narcoleptics, pharmaceutical treatments are necessary for integration into mainstream social life, particularly in terms of modern work, school schedules, and family life. For those who exempt themselves from the use of pharmaceuticals, and even for some who do not, napping treatments prove only marginally effective, often requiring social sanctions for their exercise. The increased pharmaceuticalization of sleep and its disorders in American society might decrease a tolerance for napping, with the expectation that disorderly sleepers will seek pharmaceutical help rather than negotiate social allowances for napping or for flexible work schedules. With pharmaceutical companies' increased persuasion for the medicalization of sleep in American society, the choice to decline medical treatment, as in Sam's case, may become increasingly rare and perceived as dangerously antisocial. The desire for normative sleep, regardless of the intimacies involved, dangerous or benign, may become hegemonic in its demands of individuals and their mediated relationships with the institutions that constitute the spatiotemporal order of American everyday life. Like narcoleptics, who are often forced to opt in or out of social interactions on the basis of their desire for sleep, individuals with obstructive sleep apnea provide an example of the need for prosthetic therapies for the mediation of interpersonal intimacies.

Three (or More) in a Bed: Obstructive Sleep Apnea

In this section, I focus on three cases of apnics, the first being that of Dave Hargett, who interviewed in 2006 and had been the chairperson of the board of directors for the American Sleep Apnea Association since 2002 and who had decided to be an activist in regard to apnea throughout his retirement. This is followed by a discussion of a couple who are both sleep apnics, an increasingly common occurrence among bed partners. I focus here on the narratives of Rosaria and Robert Kristophsen, who exhibited different forms of sleep apnea and shared their bed space with two machines, making the scene of their marital intimacy also a scene of shared human–machine intimacies. I came to know Hargett through my year in the Chicago area, where I spent time with local patient support groups, in particular the apnea support group named AWAKE (Alert, Well, and Keeping Energetic), in which Hargett was foundational in forming, planning, and leading. The meetings of this group provide a space for the newly diagnosed and longtime apnics to discuss treatments and for knowledge about new technologies and therapeutics to be disseminated. As such, the meetings are not in the confessional idiom that Alcohol Anonymous meetings are but rather are about negotiating diagnoses and therapies. Secondarily, they serve as social networking opportunities, and it was here that Hargett introduced me to the Kristophsens because of their frequent involvement in a number of support groups throughout the Chicago area, in which they took as their focus the need to help newly diagnosed apnics come to some understanding of the normalcy of being an apnic and the abilities to negotiate with one's therapeutic treatment, especially within the context of a shared bed and as intimates.

Hargett was diagnosed as a sleep apnic after a series of social travails and health complaints. At the time of his interview, Hargett had the physiology of a classic apnic, being slightly obese with a thick neck; he had recurrent health problems, some of which were related to his apnea, but others were related to his diet and lifestyle, including high blood pressure and a history of heart attacks. In our conversation, he discussed his personal experiences as a sleep apnic and the many complications it had produced in his life, as well as the history of the American Sleep Apnea Association and his leadership of local AWAKE groups, placing his life as a disordered sleeper in the context of attempting to help produce order for other apnics in their newly diagnosed lives and search for treatments. Hargett was diagnosed with sleep apnea at the age of forty-five, in 1993, when the disorder was still relatively newly recognized, and he reported exhibiting eighty-two or more apnea events an hour, a count that qualifies as severe apnea. This means that in the course of one hour, Hargett would choke to the point of waking up eighty-two times, barely sleeping through a single entire minute. In Hargett's description of his experience, marital intimacy played a central role. Unlike the experience of most apnea patients, however, Hargett's quest for treatment was motivated not by the discomforts of his bed partner but by an article published in the *Chicago Tribune:*

> I was beginning to find myself getting up almost hourly at night to go to the washroom, to urinate. I had begun to think that—this is the summer before I got diagnosed—I probably had a bladder problem or a prostate problem, and I better go see the doctor about it. But meanwhile, in September of '93, I read an article about sleep apnea in the *Chicago Tribune*—that's when I figured out "I think the symptoms described in this are what I have." So unlike most patients, where it's the spouse that drives them to get diagnosed, I read about it, thought I recognized it in myself, asked my wife—she didn't believe it initially. It then took me a year to get my machine, partly because I just never pushed myself to go check it out.

Between the time of Hargett's self-diagnosis and eventual consultation with his primary care physician, he and his wife lived through his severe apnea. As Hargett narrated their negotiations of his sleep disorder, the spatial disruptions of their intimacies figured prominently, with his wife being displaced through her own need for sleep, first to an unused bedroom and eventually to the living room couch. This movement was framed as a result of her own inabilities to sleep through her bed partner's disrupted and disrupting sleep:

I'm curious about your wife's sleep. She must be a very deep sleeper to have slept through your apnea events.

She wasn't. She was hitting me to try and get me to turn over. Of course most of the time I didn't even know it. She was getting very disturbed for a while. . . . She had put up with my snoring and hadn't really realized the pauses in breathing. So I think [my apnea] developed over time, worse and worse. . . . The summer I was having all of these difficulties, she moved into the other bedroom, which is sort of a temporary room where we have a Murphy bed; I was uncomfortable, but it was better. Except that there was a common wall, and pretty soon she was hearing me through that wall. So for a couple months she was sleeping out here on the sofa. . . . If she fell asleep first, she could maybe get some sleep. But if I went in there and fell asleep first and was snoring . . . she would have major problems trying to fall asleep. . . . We jokingly talk about one of the symptoms that's not in the literature is the bruised ribs from a spouse's elbow: "Wake up, honey, roll over."

What eventually took Hargett from his state of "sleep apathy" to being diagnosed with sleep apnea and eventual treatment was not the disrupted intimacies of the marital bedroom but rather a worsening of his performance at work, a series of minor automobile accidents, and concern with his overall health.

Being diagnosed with sleep apnea in the 1990s offered few treatments (primarily CPAP machines) and only a very limited variety of machines and masks. By the turn of the twenty-first century, the machinery involved with the treatment of sleep apnea had broadly developed, through the interests of both patients and designers and manufacturers, the latter seeing sleep apnea and its accoutrements as a site for possible sleep consumerism. The question of treatment and its successes structures the recounting of Hargett's experience as a sleep apnea patient, and it brings together the concerns of bedroom intimacy with compliance and patient education in his narrative of his eventual diagnosis:

I was in [my primary care doctor's] office at 3:30. At 7:30 that night, there was a home care company rep here with *a* machine and *a* mask, no choice. And she tried to explain things, and of course when she left and me and my wife are watching her get in the car we thought, "Oh my god, what did she say? Can I use this thing?" 'Cause when she was showing me how to use it, she had the mask in her hands, and she's bringing it to me, and she has it two to three inches from my face, and she powers on the air. Well, that's not a good way to introduce you; that's like the proverbial dog sticking his head out the car window, ears and hair flopping, because you get this rush of air, and it takes your breath away, and you think, "Oh my god, there's no way I can sleep with this damn thing." And so she left, and we're so panicked, and we're not sure what's going on, so we said let's both call in and leave word at work that we're not coming in tomorrow, 'cause we're not sure what's going to be going on tonight. So we finally got brave enough to go to bed, and I put the mask on, lay down, and got comfortable, and promptly fell asleep for five hours. So I knew there was something special going on when I didn't have to get up every hour to go to the bathroom. Now, my wife laid there those same five hours with her hand on my chest because I wasn't snoring, and she wanted to be sure my chest was going up and down to make sure I was still breathing. I was more rested the next day than she was.

Hargett's narrative is well honed, a result of his role as a support group leader and a public figure in the dissemination of knowledge about sleep apnea and its treatments. By bringing together his experience both as a patient during a less technologically expansive era and as a husband and bed partner, Hargett simultaneously evidenced his grandfatherly role among sleep apnics (despite his comparatively young age) and his everyman-ness as a disrupted and disrupting intimate and spouse. This was supplemented with an acknowledgment on his part for the need to become intimate both with his therapeutic technology and with himself as a more orderly sleeper.

On the Saturday, following his use of the CPAP machine, a day on which he would normally feel lethargic and nap, he finally realized the success of his treatment:

At 11:00 that night it dawns on me: "Hey dummy, you haven't felt sleepy today. You didn't take your two-hour nap; you kept going all day long." I literally walked out into the hallway and banged my head against the wall and said, "You dumb son of a bitch, this thing really works." For me I saw a major difference in how I felt, not so much how I really, really felt, but in the need for sleep, for that nap. So, of course, I didn't nap on Sunday; I saw a difference in four days.

The line I use with most patients in my support groups these days is, "I saw a difference in four days; some people, it takes four weeks; some four months; and I actually know one person who struggled with CPAP for four years."

Throughout Hargett's narrative of his illness and treatment, he deployed thematic refrains about the beneficial effects of CPAP and BiPAP machines, the necessity for compliance on the part of patients, and the need for the patience of bed partners—all calls for reconfigured intimacies across bodies, between bodies and machines, and between individuals and their institutional roles. This is most evident in the following apocryphal story, related to Hargett by a representative from one of the sleep apnea technology companies, which Hargett retold in our conversation and at support group meetings:

> This seventy-year-old man—I'll call him Frank, his wife Mary—was diagnosed with apnea and prescribed a CPAP. So the home care rep comes in; she asked Frank to have Mary there because she always likes to show both partners how the machine worked and describe it. Sometimes the apnea patient, whoever it is, doesn't always remember everything right or just doesn't remember it at all, so it's always nice to have both partners there. And the lady [Mary] says, "Oh, all right, I'll be there"; she was real grumbly, and she said, "I'll help him get started on this thing, but when he's started and he's feeling better, I'm out of there. This is not the man I married fifty years ago; he's nasty, he's irritable, he's a terrible person to me these days. I can't stand him anymore; I'm leaving, and we're getting a divorce." . . . Two weeks later the tech calls back to Mary and says, "How are things going?" And she says to the tech, "Frank just made us reservations at a local romantic hotel"—she was giddy.

Throughout Hargett's narratives—his own and those he attributed to others—intimacy between partners and the need to adapt themselves and their bed partners to machines, to the noises and benefits of those machines, bring together diverse experiences of disorder. Complex relationships between human bodies and desires for sleep are brought into accord with the rhythms of therapeutic machines, and this intimacy is founded on the burgeoning capacity of the newly ordered, sleeping body and its prosthesis. There is always the possibility of noncompliance, that those involved will be unable to forge this intimacy, and there is also the possibility of even more complex therapeutic intimacies, involving more bodies and machines.

The success of apnea treatment is equally evident in the case of Rosaria and Robert Kristophsen, who, as mentioned above, were both apnics and found a reconfigured intimacy at the heart of their parallel diagnoses as disordered sleepers. Both were relatively young at the time of interview: Robert was thirty-five, and Rosaria was twenty-nine. They had been diagnosed one and three years before, respectively. Unlike Hargett, Robert and Rosaria had less stereotypical bodies for apnics, although Rosaria's neck was very short, which she cited as being one of the causes of her apnea. Robert, on the other hand, had no stereotypical associations with apnics, which he saw as a stumbling block to his diagnosis and eventual treatment; clinicians had thought that apnea seemed an unlikely cause of his poor sleep. What had led Rosaria to seek treatment was her father's diagnosis as a sleep apnic: his primary care physician had asked him if he had any children and if they had symptoms similar to his own. He noticed that when Rosaria would come home from college, she would spend much of her time napping on the living room couch and loudly snoring. He consulted with her, and after she took an online survey to identify symptoms, Rosaria came to understand her chronic sleep deprivation and sudden weight gain of seventy pounds as the result of severe sleep apnea. Similarly Robert's father had been diagnosed as an apnic, and while Robert was visiting his parents his mother noticed that he was a loud snorer. As Rosaria narrated their relationship, their shared lives as disordered sleepers were placed within the context of their intimacy as bed partners:

> We met in May of '02, and Robert used to always complain that he could hear me snoring, and I was constantly tired—more so than him; he was a lot more energetic then. I was constantly tired, I'd wake up with extreme headaches, and one headache—I couldn't even go to work it was so bad. . . . We started dating in August of '02, and like I said, I was diagnosed and being treated by December of '03. I do think that exterior conditions of sleep apnea were affecting me more than they were affecting him.

> *Were you aware of symptoms in Robert after you got treated?*

> Yes, especially a few times I got worried because he choked and coughed. He used to fall asleep right away . . . and he'd start chok-

ing. . . . He's snoring and stuff, and I didn't really realize that, because he usually came to bed after me, and by that time I was with my CPAP machine and I didn't hear him.

Their individual disorders served as a mechanism for their support of each other and acted as a foundation for their intersubjective understandings of each other as disordered sleepers, bed partners, and intimates, in some cases through the machines that both Rosaria and Robert relied upon for normal sleep. Their desire for sleep both molded and was molded by their desire as intimates. But their intimate relationships with their treatments allowed for new intimacies to unfold.

The Kristophsens' shared burden of sleep apnea diagnoses and their ability to negotiate each other and their individual health concerns were what endeared them to Hargett. The Kristophsens would routinely attend local AWAKE support groups to evidence how a couple could successfully found new intimacies despite sharing their bed with medical prostheses. As Robert related in our conversation:

> We support each other. . . . We were kind of joking [when we were first diagnosed] that we were kind of like two elephants with [air] tubes, and someone said, "Well, can't you have the things [CPAP machines] with like a splitter?" and I was like, "No, she's at one pressure, and I'm on a different one, so we have our separate machines." We have to have, obviously, a nightstand on either side of the bed.

Rosaria narrated their shared duties with their machines, both in terms of their daily care for each other and with a view to their continued marriage and apnea symptoms:

> I take care of cleaning the masks and our chambers and our hoses; he fills our chambers at night, and I empty them in the morning. So it's pretty much routine. . . . The only thing is that I don't know how to get Robert's chamber open to empty the water in it. I need to learn how to do that so that if he's ever in the hospital or something and can't do it himself, they say the spouse should know how to set up the equipment and put the mask on.

Rather than perceive their disorders as a site for tension between them or as individual burdens, the Kristophsens accepted them as a means to express their intimate relations with each other, as mediated through their relationships with each other's therapeutic machines. As in the case of Hargett and his wife and the narratives of other apnics that Hargett incorporated into his discussions of ongoing therapy for sleep apnea, the Kristophsens' collective treatment depended upon their intimacy with each other and with each other's therapeutic prosthesis. These relationships, as Rosaria made explicit, were often direct, such as in her cleaning and maintenance of both of their machines and in her learning how to operate Robert's machine as well as her own.

In the following, Robert and Rosaria reflected on the way sleep disorders affected their social relationships but also offered the possibility of reconfiguring those relationships:

> ROBERT KRISTOPHSEN: The good thing is that we support each other and understand what each other is going through. We have different masks and different pressures, but . . . we have that support for each other. It's not a curse.

> ROSARIA KRISTOPHSEN: Plus, I think, too, with our fathers having it, we realize how this is affecting our whole family and how it affected their spouses and us, and—everyone's involved in this. Whether our mothers have it or not, they're still involved.

This intimacy between Rosaria and Robert was achieved through their mutual understanding of each other's health complaints and the dependencies that such disordered sleep enforced in their use of technological prostheses. Apnics offer an example of bodies interacting with and extending themselves through prosthetic therapies, whereas other disordered sleepers offer cases of molecular therapies, pharmaceutical and chemical prostheses that alter human capacities through more discrete means, affecting bodies and their interactions with the world.

The Delicate Balance of Insomnia

At the time of our interview, Betsy was a white woman in her early fifties, having retired early from her career partly as a result of irresolvable insomnia. She described the impacts of her insomnia on her work life:

> It made work difficult. It made me take drugs that I didn't want to take. But I couldn't manage my insomnia without taking drugs. I've had very demanding jobs, and I needed to be sure that I could function. Some people, I guess, can function well with little sleep, but I

just don't function well with little sleep. . . . And as I look back, I think I would have made different decisions about work . . . and I probably would have done better with a less demanding job. I don't think it's good for you; I think it messes with your brain. I used to think I was unusual, but now I think I'm quite typical.

Betsy's case was exceptional, not as an insomniac, but as an insomniac who could jettison the obligations of work early in life: because of economic support from her husband and having already ensured that her children had made their way through school and established themselves in careers and with families of their own, Betsy had been able to muster out of the workforce. She explained her life outside work specifically as a reaction to the pharmaceutical ordering of her life that the obligations of working forced upon her:

Therapeutically what have you tried?

Oh, lots of drugs. Lots and lots of drugs. Everything from benzos to Xanax, antidepressants, and all the tricyclics. . . . Let's see, I've even taken muscle relaxers mixed with other drugs. And they're effective for a while, and then they all wear off. I've taken them for years, and when one stalls out I go back to the doctor and get a new drug. And drinking [alcohol] helps. But I know better than to do much of any of them. . . . I would get dependent on all of this stuff, and when it came time to change I would need to wean myself off of it.

How are you treating it now?

I am off all drugs. Since I moved here about five years ago, I am no longer working. I retired early. And so I have a lot more flexibility of time, I don't have to get up early in the morning, and I don't have the stress of a job. . . . I've been messing around with my insomnia since my life has changed a lot. And I just decided, about a year ago, to go off all drugs. Nothing seemed to be working. I mean, nothing worked for a long time without side effects. . . . It took about six months to get totally weaned off everything, so now I just practice good sleep hygiene.

This ability to except oneself from the obligations of work and the chemical dependencies that those obligations can generate among insomniacs is uncommon, and most insomniacs struggle to regulate their heterodox sleeping patterns in order to be fully integrated into the broader spatiotemporal order of contemporary American society. Betsy's ability to treat her insomnia without pharmaceuticals and caffeine, but rather through the social management of her everyday life, is remarkable; for Betsy, living with insomnia and work meant being chronically sleep deprived, and attempting to manage the social and biological chaos produced by her sleeplessness led inevitably to investing herself fully in the everyday intimacies of contemporary social life, especially work, through reliance upon chemicals. Entrenching oneself in the everyday orders of desire and intimacy that work often necessitates only helps to exacerbate the unruliness of sleep disorders. Chemicals offer discrete forms of therapies, acting upon the biologies of those who employ them, in turn rendering emergent desires and intimacies and altering those extant intimacies and desires that make up individuals' everyday lives. This is no more apparent than in the case of REM behavior disorder.

REM Behavior Disorder: Inverted Intimacies

In the case of REM behavior disorder (RBD), new intimacies can be founded between bed partners, but the phenomenology of the disorder is psychologically and physiologically disruptive, and it produces crises in relationships because of its extreme symptomatic expressions. These effects are in part the result of the expectations of what such severely disordered sleep apparently explains about the intimacy of bed partners. What is involved in the negotiations of the cause of these bedtime behaviors is a clinical placement of RBD, and sleep disorders more generally, firmly in the realm of the physiological rather than the psychological; RBD as a boundary diagnosis is the site of contestation between sleep specialists and sleepers, because physicians attempt to explain its pharmaceutical fix of physiological problems despite what appears to patients (and sometimes to other researchers) as a dramatic psychological disorder. The problem faced internally by the researchers and clinicians who originally diagnosed RBD is that while treatment with small doses of the drug Clonazepam is "exquisitely effective," they have no idea how or why it works, according to Mark Mahowald, the neurologist involved in the initial diagnosis of the disorder. Moreover, as explained by Carlos Schenck, the psychiatrist who defined RBD, "It's natural for a lot of people, including physicians, to conclude that [RBD] must be a psychological or psychiatric disturbance because of the phenomenology of the experiences: screaming at night, running around, punching your wife—can't get more psychological than that." In the following, I draw on my fieldwork at MSDC, an interview with Schenck and his *Paradox Lost* (a collection of cases of parasomniacs), and the case of Mel Abel, a man diagnosed with RBD who was foundational in the concretization of the nosologic category.[7]

RBD bodies relentlessly attack bed partners and their sleeping environments, and to rectify this, novel contraptions have often been employed. In the following, Mel Abel describes his attempts at physical restraints:

> I started tying myself to the bed with a rope . . . when I started get-
> ting the bad dreams. . . . I got a piece of clothesline rope that I tied
> up at the end of the bed. . . . I got a loop on [one] end where I got
> a heavy belt going through . . . and put this belt around me, because
> I do turn and flip around a lot during the night.[8]

Other examples of makeshift remedies are similar. The most dramatic case described to me by Schenck is that of a couple who had installed a piece of Plexiglas between two single beds, so that they could sleep next to one another without the wife bearing the brunt of her husband's RBD aggressions. A desire to retain the scene of marital connection, of intimacy, has motivated these experimentations with sleeping space, and when they've proved ineffective, husbands and wives have finally separated into different spaces altogether.

Because of the age of onset in RBD, the first generation of patients diagnosed in the early 1980s were born in the 1930s and earlier and had been married for decades by the time of their search for medical expertise to fix their intimacy problems. In this marital context, the violence of the husband's sleeping behavior was subject to interpretation that assumed a psychological cause, despite expert protestations to the contrary. After years of what was often described by patients and their bed partners as relative bliss, disrupted sleep intervened to cause bedroom and marital strife, often in counterpoint to a waking life that retained earlier intimacies. The staff at MSDC came to largely accept that these RBD patients had serene daytime personalities, best captured in Schenck's introduction to *Paradox Lost*:

Patients with RBD usually have calm and pleasant personalities, and do not display irritability or anger while being awake. Although this has not been rigorously proven in a scientific study, it is a generally accepted maxim by most clinicians familiar with RBD. . . . Only in RBD is this personality profile found so prominently. So what is it about RBD that the people who have it are almost always so pleasant and mellow?[9]

Scheck's question is a perverse one, mixing evolution, biology, society, and psychology into a strange causal chemistry. It is in this matrix of causal possibilities that RBD has developed, and this matrix continues to lend veracity to psychosomatic explanations for a disease that has been increasingly evidenced as physiological and without psychological components. In *Paradox Lost,* husbands recount their dreams and violent behaviors, and wives air their attempts to grapple with their troubled knowledge of their partner, who throughout the day is the polar opposite of his nocturnal self. In one case, a wife explained her husband's temperament: "He's so foreign to anger. Dumb things will make one angry, like other drivers or something, but I always say he just lets problems come and roll right off him. That's why these dreams are so shocking to me."[10] She explained further: "At first I blamed it on the book [he was] reading. I blamed it on that and then I wondered if it was coffee. He was drinking coffee at night, but his stopping this didn't work. Then he said it was not enough sex," to which the husband explained that whenever he and his wife had sex he "never had a bad dream."[11] Throughout these intimate deliberations, partners negotiate being with one another against the backdrop of psychological and physiological distress. Explanations that stressed the former over the latter were ruined when psychiatry and neurology intervened, producing a pharmaceutical treatment that relied in no part upon psychological explanations for RBD.

The remedy for RBD is relatively straightforward, involving small doses of clonazepam, which is marketed as Klonopin in the United States and Rivotril elsewhere. Clonazepam is primarily used as an anticonvulsant, and Schenck and Mahowald, as they described to me, guessed about its use in attempting to suppress the dream enactment behaviors pharmacologically; they presumed that as an anticonvulsant it would trigger the loss of muscle tone, a loss that RBD bodies don't experience during REM sleep. In other words, while most sleepers are paralyzed during

REM sleep, preventing them from acting out their dreams, sleepers with RBD are not, but the clonazepam works to produce this paralysis pharmaceutically. Clonazepam works in about 90 percent of RBD bodies and has the unexplained side effect of also curtailing violent dreaming. Some other treatments have included the use of melatonin and stress-reduction techniques, both of which have had dubious results, and dopaminergic pharmaceutical regimes, which are effective for most who exhibit RBD symptoms.

What I can only hint at here are the difficulties that a physiological diagnosis raises for couples who involve an individual with RBD symptoms and have explained the presence of his or her unruly behavior as necessarily psychological. Because popular explanations lend themselves to understanding RBD as some eruption of suppressed hostility, a pharmaceutical fix can be a disappointment; it can disrupt the intimacy developed from being with a partner with extreme sleep disruptions. This is especially the case when family life has been reorganized around the understanding of RBD as sublimated aggression, including, as mentioned above, the use of paraphernalia to maintain the marital bedroom scene. This can be the case despite the crises provoked by such interpretations of disordered sleep and machinations around it, which may have to do with this historical moment in the transition between psychological explanations and pharmaceutical treatments for the disorder. One element of this is the tension between the permanent fix that psychiatric therapy might provide and the chemical dependencies that contemporary medicine offers to RBD bodies: the condition never resolves itself and instead must be medicated against daily. Another element of this tension is that instead of the disordered sleep being understood as shared only by the couple, as necessarily a product of intimacy, it is conceptualized as a manifold concern; at the end of the day, the disordered sleeper's problem is her or his own to medicate or not, but the symptoms of it are shared by any proximate cosleeper or bedroom furniture and become integral to all social and biological relationships. Thus, the intimacy shared between disordered sleeper and bed partner is replaced with a pharmaceutical intimacy between the disordered sleeper and his or her medication, and this pharmaceutical intimacy is mediated by a clinician and is predicated on the need for medical intervention. But there may be a middle ground, an interpretive space of desire and intimacy where an ambiguous relationship between physiology and psychology, chemistry and cohabitation, might linger, as described by Schenck in our discussion of a case of a sleepwalker who was abused in her childhood:

> For some [parasomniacs], yes, certainly there is a causal relationship [between abuse and the manifestation of a sleep disorder], but you still have to be genetically predisposed. . . . I saw a patient very recently that was fascinating. . . . She started sleepwalking—just uncomplicated sleepwalking—at the age of three that lasted pretty much weekly for a number of years. Then her father sexually assaulted her—actually began to sexually assault her—at the age of ten. Shortly thereafter she developed sleep terrors. The point is that she already had disordered arousal and a non-REM sleep parasomnia beginning at age three with sleepwalking. So the sleep terrors unmasked what was biologically there. Maybe she never would have developed sleep terrors, but she already had disordered arousal before the abuse, so that's kind of a mixed picture that way. If she was not a sleepwalker, maybe she would have developed horrible nightmares out of REM sleep and never had sleep terrors. . . . You can't be too biological—that's very bad. You have to be much more global and explore the psychology of the person and the family history as well. I think more often it's a mixture of the biological and the so-called psychosocial.

So far as extant explanations of RBD go, there are no "psychosocial" causes of the disorder; I offer this middle ground as an intimate space that might be deployed more globally to understand sleep and the relationships that it produces beyond the physiological and pharmaceutical fixes offered by contemporary sleep medicine. This might be unsatisfying or seemingly romantic, but it is a way to bring together the various kinds and qualities of intimacy and expertise across social forms, between patients and clinicians, between patients and their bed partners, between disordered sleepers and their treatments, and between pharmaceuticals and their mediators. If the dyad of psychology–physiology is brought into these social networks, a perverse chemistry of nature and culture can be seen operating in the expertise of oneself and in the sleep of others. Objective expertise might be replaced by subjective intimacies that bind actors together across bodies, diagnoses, therapies, and the relationships they produce and that constitute the order and disorder that make sense of disrupted sleep and its daily wages. Accepting sleep's desirous and intimate logics allows for treatments that might be primarily physiological or technological and also allows for recognition of the social and psychological fixes that are required to make sense of sleep disorders such as RBD and sleep apnea in intimate relationships. And, finally, such an acceptance allows for dialogical explanations to emerge between those who are experts on the sleep of others and those who are experts on their own disordered sleep and the disordered sleep of bed partners, thereby producing intimacies that provide means for thinking beyond pharmaceutical, technological, and nosologic fixations of sleep and its disorders.

Desire and Disorder, Order and Intimacy

Across kinds of sleep disorders, individuals confront personal desires for sleep that chafe against institutional orders. In some cases, these other desires for sleep can be resolved through medical interventions. In other cases, medical interventions prove intolerable for individuals and their families. In any case, individuals form their own orders, everyday lives that meet their desires and intimacies. The contexts of these everyday lives are the institutions that shape and are shaped by these same desires and intimacies. Our desires for sleep serve as a mechanism to connect these many institutions, which find their logic in appeals to our human nature. This natural basis for both individual behaviors and institutional orders makes everyday desires and intimacies appear to have long-standing, if not eternal, significance. Conceiving of our desires for sleep as founded in prehistorical, precultural nature is an appeal to primordial logic, a justification for contemporary formations that makes claims to a basis in reality despite being fictional. From this basis, individuals come to understand their alternative orders as *disorders*. Some institutions develop to mitigate these disorders, offering individuals other orders through which they might organize their lives. Simultaneously, these institutions offer yet other orders for everyday life and our desires and intimacies. Whether or not these other orders will succeed has as much to do with the ways that individuals and other institutions invest in them as with

5

· · · · · · · · · · · · · · · · ·

Now I Lay Me Down to Sleep

Children's Sleep and the Rise of the Solitary Sleeper

C HILDREN'S LITERATURE THAT TAKES BEDTIME AS ITS CENTRAL
narrative concern is expansive;[1] among the dominant themes in the
stories are naturalizing solitary sleep and rendering the bed and bedtime
as anxiety free. In so doing, the tacit objective of much of children's bedtime literature is forming normative spatiotemporal desires: early birds
who catch worms and who internalize the need and means for sleep as
solitary, moral processes. The progenitor of these contemporary texts
is *Goodnight Moon,*[2] which shows the bedtime ritual of a young rabbit
protagonist who says "goodnight" to his surroundings. The child rabbit
has a bedroom filled with objects mundane and otherwise, including a
telephone, a red balloon, an open fireplace, kittens, mittens, a mouse,
a toy house, a comb and brush, and "a bowl full of mush," as well as "a
quiet old lady who was whispering 'hush.'" Although these various objects might conspire to keep the child awake, especially the two kittens
and the young mouse, the child's bedtime ritual of saying "goodnight"
to each of these potential interferences seems to ease him to sleep. The
rabbit protagonist of *Goodnight Moon*—in a perversion of Adam's task
to name the animals of the world—finds peace through the process of
naming his environment and setting it to sleep, including "noises everywhere." The young rabbit protagonist of *Goodnight Moon* serves as a
model sleeper, one at peace with his environment, cluttered as it might

be with distractions, and ready for solitary sleep, which he has invoked through his ritual.

In this chapter, I focus on three origin stories, three sites where normative desires for sleep are established. The first is children's sleep and the practice of parents cosleeping with their young children, which is a recurrent concern among parents, pediatricians, psychologists, and scientists. The second is children's literature, which takes as its object the formation of normative desire for sleep: solitary, consolidated, and immune from distraction. I trace these themes through a handful of representative stories, suggesting that children's literature is less benign than it may at first appear. Children's literature, whether it is effective or not, provides a glimpse into how and when American desires for sleep are shaped. In the final section of this chapter, I turn to the everyday institutional rhythms of American society and how they instill a desire for society in individuals, a counterpoint to the desire for sleep. Taken together, these three origins of desire help frame the experience of individuals as disordered sleepers and suggest the individual and institutional complexity of disorder and desire.

David and Karen are the parents of three boys. By their own accounts, this couple have always been model sleepers, sleeping about eight hours each night. The time of going to bed and getting up in the morning might vary, but they are reliable, steady sleepers. Sometimes David might stay up late working, writing into the night, but he could easily catch up on his sleep the following day. Sometimes one of them might have a sleepless night because of stress or some unknown cause, but it would be fleeting, and they would return to a regular schedule soon enough. Through Karen's first pregnancy, they both slept fine. She might have had an uncomfortable night every once in a while, especially as her pregnancy advanced, but she made up for it with a nap the following day. When their first son arrived, they shared their bed with him, and this was not a contentious decision.

Josiah slept between them or sometimes to one side of one of his parents. If he awakened in the middle of the night, hungry, Karen would roll over to expose her breast and feed him. She would sleep through most of his suckling and eventually Josiah would return to sleep as well. They had a king-sized bed, and they had removed the frame so that there was no danger of Josiah falling out of bed; if he tumbled out, he would fall only inches to the floor. And the king-sized bed seemed like more than enough room for the three of them. When Obadiah was born two years later, they just made room for him.

Their home being a ranch with a finished basement, the bedroom was located just beyond the living room. The boys would wander back whenever they wanted a nap, or if one of their parents was napping, the boys might join him or her or rouse them if they needed something. At times, one family member's nap would become a family nap; other times, one of the children would bother the napper, ensuring no rest for anyone. But, overall, the situation allowed for a variable pattern of sleep and wakefulness. The third brother's birth, about two years after Obadiah's, coincided with Josiah's interests in having his own room. David and Karen set up bunk beds in the bedroom across the hall from the master bedroom, intending them for Obadiah and Josiah to share. Even so, the boys would sometimes sleep with their parents and new little brother. Neither David nor Karen was a restless sleeper; they didn't thrash about or roll over incessantly. They weren't afraid of smothering one of the boys. And there were other rooms in the house for their sexual purposes; they didn't feel confined to their bedroom. It might not have been an ideal arrangement for everyone, but it worked for their family.

Cosleeping has recently reemerged as a topic of parenting conversations and sleep science. In 1999, the U.S. Consumer Product Safety Commission (CPSC) issued a warning that "64 deaths each year [result] from suffocation and strangulation" during bed sharing between parents and their babies. According to CPSC chairman Ann Brown, "The only safe place for babies is in a crib that meets current safety standards and has a firm, tight-fitting mattress. Place babies to sleep on their backs and remove all soft bedding and pillow-like items from the crib."[3] Of the 515 deaths investigated in the study, 394 were attributed to the child being entrapped between the mattress of the bed and bed fixtures, adjacent walls, or furniture or to suffocation related to sleeping on a waterbed mattress. Another 121 deaths were attributed to a parent or sibling "rolling on top of or against baby while sleeping." In 2005, the American Academy of Pediatrics (AAP) bolstered the CPSC's recommendation with the suggestion that "routine pacifier use" while sleeping might reduce the rate of sudden infant death syndrome (SIDS), otherwise known as "crib death."

The AAP went on to explicitly advocate a ban on babies sleeping with their parents, suggesting that this ban would also lead to the reduction of SIDS.[4] Although there are some sixty-four deaths each year related to sleeping in a "family bed" or cosleeping with parents, some twenty-five hundred deaths each year occur while children sleep alone in their cribs, often attributed to SIDS. Being a sleeping baby is necessarily risky; blankets, pillows, furniture, and mattresses all pose dangers, as does sleeping position. How that risk is conceived by parents, pediatricians, scientists, and consumer advocates has as much to do with expectations of the lone sleeper as it does with the science of sleep.

On the other side of this debate, which is advocacy for the "family bed," are parenting experts who focus on "natural" and attachment-parenting strategies and scientists concerned with reducing SIDS rates. Despite the warnings of the CPSC, Elizabeth Pantley, author of *The No-Cry Sleep Solution,* suggests:

> Infants should be placed between their mother and the wall or guardrail. Fathers, siblings, grandparents, and baby-sitters don't have the same instinctual awareness of a baby's location as mothers do. Mothers, pay attention to your own sensitivity to Baby. Your little one should be able to awaken you with minimum movement or noise—often even a sniff or snort is enough to wake a baby's mother. If you find that you are such a deep sleeper and you only wake up when your baby lets out a loud cry, you should seriously consider moving Baby out of your bed, perhaps to a cradle or crib near your bedside.[5]

Pantley goes on to suggest that other detrimental sleeping situations include parents who drink alcohol, parents who "have used any drugs or medications," those who are sleep deprived, especially sound sleepers, and those who are large. Although Pantley is an advocate of parenting based on "gentle and loving" techniques, how many parents can likely claim to meet her recommended conditions for bed sharing, considering her prohibitions? Although Pantley believes in innate maternal instinct, the barriers to "gentle and loving" parenting are many and social, from pharmaceutical use to obesity.[6] Or what many might accept as the inevitable sleep deprivation of new parents. James McKenna, a biological anthropologist at Notre Dame University, argues that cosleeping reduces the risk of SIDS and, more important, increases the total time of parents' sleep.[7] Whereas Pantley, like many authors of parenting guides, bases her argument on her own experiences and anecdotal data, McKenna has conducted sustained scientific research on cosleeping, often under the rubric of SIDS prevention. Americans' continued resistance to his findings as parenting advice has as much to do with our cultural expectations of normal sleep (who sleeps with whom and when and where sleep occurs) as it has with what kinds of individuals our dominant ordering of sleep is intended to produce.

American expectations of children's sleep is predicated on assumptions about orderly, consolidated nightly rest. At the MSDC, a month would rarely pass without parents seeking medical help from the staff for an unruly, young sleeper. At a staff lunch, one of the administrative assistants, charged with making appointments for potential patients, related that a mother had called the clinic seeking help for her and her partner's three-month-old child. The problem? The mother was afraid the child was an "insomniac" because it failed to sleep through the night. General laughter ensued among the staff, and the pediatricians joked about the parenting guides the parents must be using to think that a child of three months could be an insomniac. "It has to be Ferber or Spock," Dr. Pym exclaimed, to which Dr. Richards asked if parents still read Ferber. "Of course," Dr. MacTaggert replied, "why else do you think we see so many sleep-deprived parents?" Parents expect to return to their normal sleeping patterns as quickly as possible, complicated by the amount of time they receive as maternity and paternity leave; they must return to the orderly patterns of everyday life or risk disorderly sleep, and possibly medication. One of Nathaniel Kleitman's earliest experiments was charting the consolidation of his daughter's sleep (see Figure 1).[8] Consolidation was Kleitman's assumption and goal, and his early work on children's sleep provides a basis for later conceptions of sleep, but Kleitman fails to describe the experimental conditions under which observation occurred, substituting his parenting for scientific objectivity. Children's sleep, like that of their parents, is inextricable from the spatiotemporal ordering of American everyday life, and although recurrent debates focus on *how* children sleep, all parenting and scientific research points to the eventual isolation of the sleeper as an individual. This is especially apparent in children's literature about sleep.

In *Goodnight Max,* the young rabbit Max struggles to go to bed

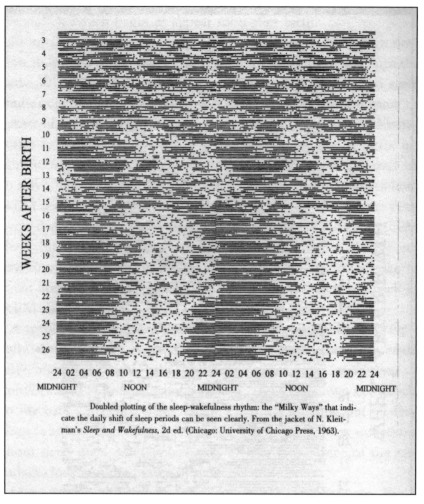

Doubled plotting of the sleep-wakefulness rhythm: the "Milky Ways" that indicate the daily shift of sleep periods can be seen clearly. From the jacket of N. Kleitman's *Sleep and Wakefulness*, 2d ed. (Chicago: University of Chicago Press, 1963).

FIGURE 1. In one of his experiments, Nathaniel Kleitman recorded his daughter's sleep from the third through the twenty-sixth week of life. Her sleep begins very erratically and ends more consolidated, although she continues to sleep in a polyphasic fashion. Reproduced in William C. Dement and Christopher Vaughn, *The Promise of Sleep* (New York: Delacorte Press, 1999), 110.

on time.[9] Ruby, his older sister, attempts to help him by minimizing his distractions, first helping to mop up a spilled glass of water, then putting new pajamas on Max, only to find they have crumbs in them. The next set of pajamas has loose candy attached to them, which is removed only for Max to be distracted by his tick-tocking alarm clock. A noisy bug, rustling curtains, a smelly sock, the light of the moon all require Ruby's intervention, and it is only when Max has his teddy bear in bed with him and a new set of pajamas that he is finally able to get to sleep. In his bedtime frenzy, Max embodies the most unruly of contemporary children, even though none of his distractions is properly technological; rather, Max is a child in search of distraction, a child who resists bedtime through whatever means are available to him. Moreover, the text includes material representations of Max's distractions: many of the pages have different textures, fabrics, and pop-up-style mechanisms for children to toy with. These distractions might tire a child out; they may equally overstimulate a baby to the point of having difficulty sleeping. *Goodnight Max* works to make children unready for bed and, as such, serves as a counterexample for much of bedtime-related children's literature, which more often attempts to prepare children for bed through naturalizing American sleep patterns and bedtime and everyday expectations.

The naturalness of American solitary sleep patterns is exemplified in *Time for Bed* by projecting consolidated nightly sleep onto other members of the animal kingdom.[10] The text surveys the animal kingdom, from mice and geese to fish and snakes, from birds and horses to bees and dogs, ending with a human child abed and being tucked in by Mother. The child is rendered perfectly androgynous, with medium-length blond hair and a soft curve to the facial features, and is dressed in a set of periwinkle pajamas, sharing in her or his blankness a means for interpellation not unlike the rabbit protagonist of *Goodnight Moon,* even though the child is also perfectly Aryan, a dubious choice for a child to represent all of humanity. Like the protagonist of *Goodnight Moon,* the child finds herself in a bed alone, a bed large enough to be intended for a human adult. Her mother, who is present in the next to last panel, is leaving the child—who bears a rather dismal visage—to sleep alone, with an invocation: "The stars on high are shining bright—Sweet dreams, my darling, sleep well . . . good night!"[11] The human child, sleeping alone, stands in contrast to every other animal in the book, each of which is presented as sleeping in close proximity to its parent, presumably its mother. The human mother's language, making refrain to a universal indicator of bedtime (the presence of stars in the sky), is shared with that of the other animal mothers, who are also settling their children to bed and citing the coming of darkness, as well as remarking on the desire for physical closeness and explaining that "the whole wide world is going to sleep." Thus, although a principal human sleeping pattern is set apart from those of other members of the animal kingdom—namely, that humans sleep alone—ideal human sleeping times are naturalized by showing that all animals retire at the same time of day, even though some of the species presented in the book are nocturnal animals: universalism when it serves to make nature harmonious with cultural expectations and particularism when it provides a means to understand human exceptionality and cultural expectations. But *Time for Bed* is surely intended to bring peace to a child abed alone, a ritual not unlike that enacted in *Goodnight Moon* and long-standing lullabies.

In an explicit attempt to rehabilitate an otherwise troubling bedtime rhyme, Sylvia Long rewrote "Hush Little Baby" to be more comforting to children.[12] Long attempts to replace the problems she associates with modern life with a more "natural" way of being in the world. She writes in explanation:

> One of the songs that has bothered me as an adult is the original version of "Hush Little Baby." In it, a mama offers her baby comfort by promising to buy him or her all sorts of things (a mockingbird, a diamond ring, a horse and cart, etc.). It seems much healthier to encourage children to find comfort in the natural things around them and the warmth of a mother's love.[13]

The original "Hush Little Baby" may very well breed consumerism, with its strange itinerary of items the parent is willing to buy to ease the child to sleep, none of which would normally be associated with sleep (things like stuffed animals, blankets, and other security objects). But Long's desire to replace this consumerist drive with a harmonious relationship with the natural world fails to bear out in her revision of the rhyme. Although the mother in the story generally relies upon the "natural" world to ease her child before bed, she might be more properly thought of as providing her child with a series of bedtime *distractions,* not wholly removed from Max's bedtime plight. The mother first offers to show her child a hummingbird and the setting sun but then moves on to an appreciation of the crickets' song from the fields; she ends with coaxing her child to sleep by, slightly alarmingly, playing the banjo and singing lullabies. Along the way, the mother also offers her child more traditional bedtime securities, namely, a cherished quilt and a teddy bear, both of which—drawing from the earlier version of "Hush Little Baby," with its series of material failures disappointing the child—potentially might fail in their soothing effects, leading the mother to her bedtime banjo playing. As with the original rhyme, it seems relevant to ask whether a child who requires so much entertainment and distraction before bed is actually sleepy or whether the onus of tiring out a child has fallen to parents in an attempt to produce normal bedtimes, whether or not the child is ready for them.

Normative ideals for bedtime structure the daily lives of children, but whether these bedtimes actually promote health or not is dependent upon the individual child and her or his social and physiological particularities. If left to their own devices, children might sleep quite differently from how they have been encultured to, a culture of sleep based in part on the work of Kleitman, who, after recording his daughter's sleep patterns throughout her youth to produce a model of normative sleep, later focused on an experiment that was based on the sleep patterns of only nineteen children and that sought to describe the process of sleep consolidation. Over the course of the twentieth century, other experiments were conducted to ascertain the natural development of children's sleep patterns, but these experiments were conducted in the space of children's homes, with the cultural expectations of the parents firmly in place, hardly making for purified scientific conditions in which to understand the sleep patterns of children.[14] But, inevitably, the findings of these studies find their way to developmental psychologists, who are cited in turn by parenting guides, thereby reifying the naturalness of children's sleep patterns despite low sample numbers and questionable experimental design. What has resulted is the production of desiring subjects: children and parents in need of chemical stimulations and depressants, all coordinated in time with the demands of school and work days and through which family obligations are understood not as a natural bond between parents and child but as an alchemical association of spatiotemporal demands, social obligations, and pharmaceutical means.

Children's books serve as a means of inculcating cultural expectations of normalcy into children. It is against the portrayal of a responsible, lone sleeper that children are invited to found their sense of self, their ideas of what makes normal, natural sleep. Bound to both the moral messages of children's books and the spatiotemporal ordering of the school day—although often only tacitly—are the spatiotemporal demands of the workday. The forging of responsible economic subjects is an implicit drive of both the school day and seemingly benign children's books through this regularization of sleep. If children find themselves later in life chafing against the spatiotemporal ordering of the everyday, or if they find themselves maturing into adults with sleep disorders, they are provided with new intimacies and desires to develop, namely, in the form of legal means of stimulation and other forms of self-medication. But because of the spatiotemporal and economic demands of American society, most adults find recourse to such stimulants a necessity in relation to the difficulty of reorganizing institutional demands. These chemical supplements to everyday life maintain dominant spatiotemporal orders in the face of other desires for sleep. The repression of these other orders is not a malicious effort but a secondary effect of the traditional understanding of the everyday and a "harmonious" relationship between humans and the natural environment.

There is no smoking gun to explain the concretization of American spatiotemporal formations in their present state and their ordering over the course of the twentieth century, no simple explanation for the emergence of the modern spatiotemporal ordering of work, school, family life, recreation, and rest. Rather, over the course of half a century, from industrialization through the rise of the middle class in the late nineteenth century, roughly 1820–70, a complex assemblage of cultural expectations, federal legal maneuvers, and attempts to simplify modern life resulted in the hegemonic spatiotemporal formation that simultaneously unified work and school times, periods of family recreation, and rest. Moreover, this unification of spatiotemporal rhythms has also worked to explain natural human behaviors like sleep and variances from such foundational understandings as disorders. There is nothing natural about a diurnal period that is institutionally concretized, because in a natural state diurnal behavior can be variously expressed: the day can be ordered differently. In contrast, American spatiotemporal formations are an invented tradition that has managed to obscure its own arbitrariness through the weaving of a simple narrative of agrarian naturalness, which appeals to the cultural logic of many Americans who expect a natural explanation for human behavior. Individuals rely on this narrative in an effort to justify their complicity in the production of society through their participation in civil activities: school, work, family life, recreation, and the rhythms and perceived inevitability of capitalism.

One persistent narrative about American time is that of our collective past, one in which we lived in a harmonious balance with nature. Take, for example, the writings of Martin Moore-Ede, a leading sleep researcher who has also published popular books on sleep. Moore-Ede accepts a primordial state of human sleep, one that has been irrevocably disrupted by industrialization, electric light, and the "twenty-four-hour society," as he discusses in his book of that title: "Before the twenty-four-hour society took hold, everyone worked and slept more or less on the same schedule."[15] He also accepts that the introduction of the twenty-four-hour society was a new spatiotemporal formation in the history of humanity and that, once it was established, we could only negotiate with it, not eradicate it. Moore-Ede writes of the "design" of the human animal, a design intended for a hunter–gatherer model of society: "Our bodies were designed to hunt by day, sleep at night, and never travel more than a few dozen kilometers from sunrise to sunset. Now we work and play at all hours, whisk off by jet to the far side of the globe, make life-or-death decisions, or place orders on foreign stock exchanges in the wee hours of the morning."[16] Rooting human society in place formulates spatiotemporal orders that are tied to daily patterns of light exposure but often ignores seasonal weather patterns and changes in the amount and timing of daylight. It also ignores the development of many human societies in extreme conditions, as in the case of northern and southern communities exposed to long periods of light and dark, such as Scandinavian, Inuit, and Eurasian societies.

This manufactured primordial nature belies the evidence that in the past there was no unifying agrarian time but, rather, diverse agrarian times throughout the United States: growing and harvesting seasons differed among geographical regions, as did sunrise and sunset, and even in adjacent regions crop and livestock differences influenced the use of daily and seasonal time. This agricultural variance existed throughout the nineteenth century, during the initial expansion of public schooling and industrial work spaces in the United States, and stabilized times of schooling were agreed upon only after the federalization of the public education system in the late 1800s. However, the reasons for choosing those times are only partly substantiated, so we are left to infer that hegemony followed federalization, not the other way around. Nature did not found dominant spatiotemporal orders of American everyday life. Rather, the everyday produced a natural explanation.

Moreover, as the early issues of the *Farmers' Almanac* evidence, any student of astronomy understands that the length of days and nights changes on a daily basis and varies widely in duration over the course of a year, quite at odds with any normative ideal of human sleep that depends upon eight consolidated hours that remain stable over the year's course while following the shifting of dusk and dawn throughout. It seems that the agrarian time that had the most influence on schooltime was that of the northern United States, with its comparatively short growing season and long winter. The northern states' experience of sunlight, which is relatively scarce throughout the winter, influenced the production of both the workday and the school day. Although a substantial literature exists on the struggles to assert a national time that coordinated activity among distant spaces, the rationale for the federalization of time in the United States remains unclear, other than coordination for the sake of coordination.[17]

What is clear is that the consolidation of American time was determined by the northeastern states, the seat of national governance and economic influence, and they most likely acted in hegemonic fashion, forcing those who would interact with the government and corporate seats to adopt similar times of activity. What is also apparent is how the northern and midwestern states shaped a school and work day to fit the "Protestant ethic" and to neatly align with the expectations of Benjamin Franklin's famous "early to bed and early to rise." The production of normative space and time in American life shows how arbitrary cultural expectations can be, how forceful cultural expectations are in shaping the everyday life of those who partake of the dominant themes and forms of these expectations, and how, thereby, these expectations produce the very category of everyday life in American society. This category is inflected with spatiotemporal and normative inevitabilities legitimated through dominant cultural expectations and with an inexorability that is attributed to the powers of the state and society as well as, retroactively, to nature. The timing of society is often juxtaposed to nature's rhythms to write this fall from agrarian grace and thereby serves as a lens through which the steady, or inevitable, spatiotemporal rhythms of modern bodies can be justified.

The reiteration of social times, of spatiotemporal codes and their policing, through the governmental practices of such institutions as the workplace and school generates a feeling of society, a social habitus, although for those who abide by normative practices the feeling is a subtle one; in comparison, diurnal harmony with nature is accepted as without sensation. For those who are foreclosed from normative everyday life, the feeling of society's absence is apparent and becomes a site for rethinking one's nature through the locus of sleep. This arises from a compulsion on the part of individuals to embody society through spatiotemporal rhythms, of dwelling within the workplace, the school, and the family. American society as it is discursively produced is a refraction of cultural presuppositions of normal and productive uses of time, aligned with the expectations and obligations of American capitalism, which individuals defer to in order to understand their lives through narratives of normative cultural expectations, which are based, in turn, upon ideas of natural bodies and spatiotemporal rhythms.

While earlier theories of society often relied heavily on its spatial specificities—its borders and communities—we now know that society depends on the solidification of a spatio*temporal* regime and also on the management of the spatiotemporal fixity of bodies, of where they are and *when*. In this dependence, society is produced through the management of bodies and is embodied by individuals through their spatiotemporal placements and subjective understandings of one's normalcy, a biopolitical formation that is structured from above and below simultaneously, producing the controls and contours of American society. This is to say, the governmental solidification of the school day and workday allows for both the management of bodies during the day (while at work and school) and at night (while those who abide by the normative use of time sleep, and those who do not become apparent and can be policed), as well as producing a compulsion on the part of those foreclosed from normative everyday life to feel society, to share in some desire for its embodiment.

Individuals experience sleep disorders as social disorders, because they foreclose integration into normative American everyday life. Those affected include a large portion of the population who fall into the category of being late risers, or owls. Their counterparts, larks, have no difficulty abiding by Benjamin Franklin's and William Camden's spatio-temporal prescriptions and hence have no difficulty fitting into the traditional American workplace or school, but owls often experience insomnia in that they are unable to sleep when they should to be well rested and able to work a 9 to 5 workday. While sleep disorders can be treated through medication or surgery, social disorders often cannot be adequately treated in the same fashions, and they demand that dis-ordered sleepers alter their social expectations to meet their desire for sleep—to embody a new intimacy, a new desire. As might be expected, the greatest resistance to the cure for these ailments is the individual's understanding of normative ideals and what it means to be a part of American society; society is often assumed by individuals to be more inevitable and inflexible than nature and one's desire for sleep. Through spatiotemporal regulations, sociality is shaped, and it is critical that we confront the spatiotemporal orders of society and its institutions, their apparent inflexibility, and the way individuals, through no fault of their own but because of their desire for sleep, feel foreclosed from experienc-ing the intimacies of society. To ignore these dynamics is to overlook how the rhythms of everyday life conspire to form not only society but also individual medical and social disorders that challenge the formation of sociality itself.

The formation of and recourse to a normative period of time for work, school, recreation, and family life concomitantly determine what is normal and abnormal use of time and create medical and social cate-gories of disorder, feelings of disenfranchisement, and a compulsion for embodiment—a desire for a certain kind of sleep. Dominant cultural expectations of normalcy reinforce arbitrary decisions regarding the use of space and time. Whatever the veracity of the medical evidence that supports altering spatiotemporal arrangements, such as through flexible work time and schooltime and workplace napping, culture can slow any such attempts. Ultimately, school hours are decisions of society and the state, and as such they produce the contours of society. The participation of individuals in this normative spatiotemporal formation is what makes society a sensual pleasure beneath the level of awareness, an object of de-sire. Work hours are slightly different from school hours in that specific business owners set them. But, because of the apparent need for businesses to be synchronized, any business uncoupling itself from dominant spatiotemporal orders places itself at risk. Following a similar logic, although *flextime* was a key word of the 1990s workplace, most corporate workers maintain 9 to 5 work schedules, do not telecommute, and abide by normative uses of space and time. Although the American ideal of a 9 to 5 workday is not as prevalent in the early twenty-first century as it once was, it continues to act as a normative ideal that structures the way individuals conceive of their desires. In conversation, disordered sleepers who do not work 9 to 5 days, or need to, often make recourse to the idea of being *able* to work a 9 to 5 day, as if it is something eminently desirable. They are compelled, in other words, toward the spatiotemporal order of normative society. This is not to say that all disordered sleepers are captured by this compulsion to align themselves with the space and time of society; there are surely contrarians, as in the case of Sam and Dee, but those patients who purposely remove themselves from normative everyday life are aware that they are doing so, and, for them, their desires for sleep serve as a justification for abnormal social behavior.

This may all seem a far cry from the sleep of children, bed sharing, and children's bedtime literature, but it is in these seemingly benign practices that Americans lay the basis for the society that children will inhabit throughout their lives. Training children to sleep alone, and parents to want to sleep apart from their children, lays the basis for a particular desire for sleep and wakefulness. This model of sleep is based not so much on some naturally occurring need, or even the rise of modern risks and dangers, as on a fabricated conception of what American sleep was and could be. Sleep might be arranged in various ways, both temporally and spatially, but the rigid model of dominant American society leads to specific formations of desire: individuals and institutions strive toward a normative model of time and rationalize the desire for this on arbitrary practices and cultural fictions. Normative desire facilitates the functioning of everyday spatiotemporal hegemony and is in turn formed through that very same hegemony. There may be no clear explanation of American everyday life, but a variety of benign everyday practices helps to ensure that dominant spatiotemporal ordering remains what it is and has been, for at least the last century of American social life.

6

·················

Pharmaceuticals and the Making of
Modern Bodies and Rhythms

I N 2005, SEPRACOR LAUNCHED A U.S. ADVERTISING CAMPAIGN
in the hundreds of millions of dollars to promote a sleep aid, Lunesta,
reportedly spending up to forty-three million dollars in one month. Lunesta
was developed as a commercial competitor of extant sleep aids that already had the support of health insurance companies, most of which
lacked the wherewithal to commit similar capital investments to advertising. Since most patients require some financial aid in defraying the
costs of medical treatment and pharmaceuticals, releasing new drugs,
which can be on the market for years before being accepted by medical
insurance companies as legitimate treatments, is risky business. In the
past, most pharmaceutical companies have released drugs and only after
years of little commercial attention—other than through drug representatives, who broker with physicians to prescribe their wares—do they
attempt to market their products fully, sometimes depending on a new
market to emerge, other times producing a new market for a drug. What
changed this dynamic was the success of multiple antidepressants in the
early 1990s, followed by the wild success of erectile dysfunction treatments in the late 1990s and the early years of the 2000s, both of which relied heavily on direct-to-consumer marketing, allowed by loosened Food
and Drug Administration regulations.

At the beginning of the twenty-first century, pharmaceutical companies, having come to the collective realization that humans sleep a

third of their lives, were primed to turn sleep disorders and deprivation into the next pharmaceutical blockbuster. This resulted in a number of pharmaceuticals that were formally intended for severe sleep disorders, such as narcolepsy, being recast as over-the-counter prescriptions for emergent sleep concerns, like "excessive daytime sleepiness." One of these drugs was Provigil (modafinil), which is produced and marketed by Cephalon, which hoped to achieve some success with marketing the drug as an over-the-counter, "nonaddictive" stimulant (no claims were made about it being non-habit-forming) for individuals with excessive daytime sleepiness and narcolepsy symptoms. Similarly, Jazz Pharmaceuticals and Orphan Medical transformed sodium oxybate or gamma hydroxy-butyrate (GHB, which is known more broadly in American society as a date rape drug) into a benign treatment for narcolepsy, renamed Xyrem. In this chapter, I review the strange histories of these two drugs and their possible futures. They are sympathetic drugs in that both are commonly prescribed to narcoleptics, Xyrem as a sleep-inducing drug that is also known to reduce the number of cataplexy events, and Provigil as a drug to maintain wakefulness in narcoleptics during the day. By using both drugs, narcoleptics are able to produce a facsimile of normal American sleep and wakefulness. I follow this with a discussion of caffeine, addiction, and new chemical forms of life that are predicated on a desire for stimulation, not wholly removed from the medical model of pathology offered by narcoleptics.

Cephalon's 2004 product monograph for Provigil asserts that the drug is intended "to improve wakefulness in patients with excessive daytime sleepiness associated with narcolepsy, obstructive sleep apnea/hypopnea syndrome (OSAHS), and shift work sleep disorder." With each of these sleep disorders, individuals complain regularly of sleepiness or fatigue during the day, which has been recast under the umbrella of "excessive daytime sleepiness." Provigil is justified in the case of narcolepsy and shift work sleep disorder but is questionable in the case of sleep apnea, because once patients become used to their CPAP or BiPAP machines, they often sleep more fully through the night, and complaints of sleepiness during the day are reduced. It should also be noted, however, that because of its declassification as a drug used solely for narcoleptics, Provigil has been reclassified as a medical stimulant and can now be prescribed at the whim of attending physicians for any number of drowsy conditions, from depression to chronic fatigue syndrome. Additionally, in the case of shift work sleep disorder, Provigil has managed only symptoms rather than causes, since, from its inception, shift work sleep disorder has been recognized as a problem of the workplace and its relation to the sleep patterns of workers or, rather, individual workers' inabilities to cope with the rhythms demanded by work.

Cephalon's strategy has been to elaborate both new needs for the drugs that it produces as well as more comprehensive means to include new disordered sleepers in umbrella diagnoses, such as excessive daytime sleepiness. Throughout the Provigil literature, excessive daytime sleepiness is referred to simply as excessive sleepiness, or ES, subtly broadening the times that people might complain of unwanted or unexpected sleepiness. On its Web site for patients, Cephalon states, "People with excessive sleepiness may feel as if they just don't have the energy to do the things they need to on a daily basis, such as spending time with their family or performing duties at work." Other symptoms included mental tiredness, bodily fatigue, difficulty concentrating or paying attention, and low motivation.[1] One might notice in these symptoms the very conditions of modern life, as well as life as it has been lived for centuries. Who, it is reasonable to ask, is exempt from these symptoms? Who has time and energy to spend with their family and perform duties at work? Americans are increasingly beset by everyday demands during time formerly reserved for family and recreation, meaning that when individuals have time to tend to social life outside work, they are already tired. Those "tired enough" might turn to the powers of Provigil.

To evidence Provigil's life-changing effects, Cephalon's Web site features a number of testimonials from patients, one each from individuals diagnosed with narcolepsy, sleep apnea, and shift work sleep disorder, as well as from a director of a sleep clinic. Each testimonial is relatively interchangeable, in that each describes the life of the individual previous to the medication and the individual's newfound vitality since. The following is Donna's testimony, a pharmacist and narcoleptic "who struggled with ES":

> [Before Provigil] My life was uh, basically a non life. It was basically uh working and sleeping—that's all I did. I got up on the morning that I had to work and went to work and barely made it through

the day and uh, people heard "I'm so tired" come out of my mouth probably about 50 times during the day. . . . [After Provigil] Well, since I've started taking Provigil, I don't have the excessive sleepiness so I'm able to do things like eat healthier instead of fast food all the time cause I'm too tired to cook.

Like the description of excessive sleepiness symptoms, Donna's description of her life has a certain universality: Could many Americans claim to do much more with their lives than working and sleeping? Moreover, could many Americans claim to have time to cook meals rather than eat fast food all the time? Even for those who could answer positively to these questions, there is always the possibility of having even more time for family and friends, eating, and leisure. Provigil might not be able to increase the amount of time in your life, but it may ensure that you are able to use your time fully. In so doing, Provigil, or the vitality that it provides, might become a habit. This extends beyond the possible chemical addiction that an individual might develop, an addiction that could become another form of life with its own desires, something that Cephalon is aware of but has buried both in the fine print of Provigil's fact sheet, which is made available to physicians, and in much of the patient literature:

In addition to its wakefulness-promoting effect and increased loco-motor activity in animals, in humans, Provigil, produces psycho-active and euphoric effects, alternations in mood, perception, thinking and feeling typical of other [central nervous system] stimulants. . . . Modafinil is reinforcing, as evidenced by its self-administration in monkeys previously trained to self-administer cocaine.[2]

Individuals may become habituated not only to the energy that the drug provides but also to the neurochemical effects of the drug. Untangling these two effects is impossible: the feelings ("euphoric effects, alternations in mood") are equally social and chemical, economic and biological, a potent intimate chemistry.

Equally potent is Xyrem, a drug that when taken regularly and for an extended period of time reduces the number of cataplexy events of narcoleptics and consolidates nighttime sleep. As with Provigil, the administration of Xyrem has to be justified by physicians, and, as with Cephalon, Jazz Pharmaceuticals is aware of the broader uses and desires for the drug. The physician's reference guide for Xyrem states, "By impairing primary functions like talking, eating, standing, walking or driving, [cataplexy] can prevent patents from experiencing important activities such as holding a grandchild, interviewing for a job, participating in a meeting, going to a movie, attending a party, or working out at the gym." Although not every American desires each of these events in his or her life, the universality of the events is not unlike those outlined by Cephalon for potential Provigil users. And since cataplexy affects the most basic functions of human social life (talking, eating, standing, walking, driving), what cataplectic would choose to go without the drug? Like any pharmaceutical, however, and as discussed in the Xyrem medication guide, Xyrem has a number of side effects, "including trouble breathing while asleep, confusion, abnormal thinking, depression, and loss of consciousness"; it also promotes enuresis and parasomnias, especially in those who are already predisposed to such sleep-related activities. As reported in Xyrem's expansive literature, about 10 percent of users eventually decide to forgo the drug, its negative side effects outweighing the positive benefits it provides. However, this addresses primarily the physiological side effects, and a drug like Xyrem also involves broader social considerations.

Because of the fraught social landscape that Xyrem exists within, where it is probably used illicitly as often as it's used for narcoleptics, Cephalon adopts a stern approach with patients; recall Martin, who confessed to me that he hid his various prescription medications in a locked safe, although he lived with his parents and kept his narcolepsy largely a secret from his friends. In being prescribed Xyrem, patients are expected to watch an instructional video, ostensibly presenting directions for the preparation of the two nightly doses required for efficacy. The host of the video is a middle-aged white woman with graying hair, whose affect varies between motherly and threatening. She warns the viewer: "Careful adherence to the procedures and precautions presented in this video will ensure that Xyrem will continue to be available to you and patients like you now and in the years to come." In the seven-minute video, nearly half of which belabors the preparation of the nightly doses of the medication (three to nine milligrams of Xyrem with two ounces of water, sealed in childproof containers, placed near the bed), the host makes frequent reference to legal themes:

> When controlled substances are used for medical purposes, they can provide improvements in a person's quality of life, and for the many people with debilitating diseases like yours. However, if these medications are illegally diverted from the legitimate distribution system, the nonmedical use of controlled substances can lead to public health problems. As a controlled medication, if you, with criminal intent, knowingly divert, distribute, or sell Xyrem to others, you'll be subject to criminal prosecution.

Unlike Provigil, for which Cephalon desires an unlimited field of possible prescriptions—hence the company's production of its patent replacement, Nuvigil—Xyrem is constrained by its history of abuse and its persistent potential for use on others against their will. Orphan Medical, the distributor of Xyrem, has had to develop a means to ensure that the potential for Xyrem's illicit uses was curtailed and, to that end, has established a central pharmacy and a tracking system for both patients and prescribers. This fails to account for how patients use the drug between receiving their shipments, but it ensures a chain of accountability. Unless these mechanisms dissolve, it will continue to ensure that the spread of Xyrem use is circumscribed, unless what it can be prescribed for is widened.

Whereas Provigil might allow one's life to be maximized, Xyrem has a constraining effect on patients, in that they need to structure their lives around the dose intakes. Like that of many drugs, the efficacy of Xyrem is reduced by food intake, and patients need to ensure that they have digested their evening meal before taking their first dose at bedtime. More complicated is that patients require two doses of Xyrem for it to be effective through the night. The drug has a short half-life, and the consolidated sleep that it provides for narcoleptics quickly dissipates; to reap further consolidated sleep, patients need to take a second dose of Xyrem in the middle of the night: "The second dose is often necessary for the patient to return to sleep for the second half of the night; however, the need to awaken for a second dose of sodium oxybate is usually not an inconvenience, as patients often awaken spontaneously 2–4 h after the first dose."[3] Xyrem both consolidates sleep and reduces the occurrences of cataplexy events; the former effect is spontaneous, the latter only slowly begins to take effect after an individual regularly takes the drug, sometimes taking up to two months. No one can explain the secondary function of Xyrem.

Provigil and Xyrem are often prescribed together for narcoleptics, because these two drugs combine to treat all of the primary symptoms of the disorder. But bodies tolerate the drugs differently, and while some patients respond well to one of them, at times other drugs are required for the consolidation of sleep or the promotion of wakefulness during the day. And, for some, the benefits of the drugs are outweighed by the negative side effects. Medically, narcolepsy patients who choose to live without drugs are advised that they "should live a regular life, go to bed at the same hour each night, and get up at the same time each morning. Scheduled naps or short naps just before activities demanding a high degree of attention alleviate sleepiness in most patients. The optimal frequency, duration, and time of these naps has to be established."[4] Such habitual sleep restriction ensures that sleep is spatiotemporally localized, something that can be guaranteed otherwise only through the use of pharmaceuticals. Without the powers of regulation that pharmaceuticals deploy in the bodies of their consumers, disordered sleepers need to achieve this same regulation through behavioral and habitual means: they need to impose a rhythm, with an "optimal frequency, duration, and time." If patients fear cataplexy events, they need to regulate their emotional feelings, because cataplexy events are tied to extreme emotions. This very social, very intimate concern is addressed in the product monograph for Xyrem, a publication intended for physicians:

> Substantial evidence exists suggesting that, without effective treatment, narcoleptics attempt to manage their cataplexy by controlling or suppressing their emotions (adopting a flat affect) or simply avoiding social or other situations known to precipitate attacks. As a result, narcoleptics may be mislabeled as bored, disinterested or unintelligent.

When narcoleptics choose to circumscribe the use of pharmaceuticals, they are foreclosed not just from normal times of sleep and work but also from emotional investment in social interactions as a whole. To legitimate the use of Xyrem and other narcolepsy drugs, pharmaceutical manufacturers and marketers are behooved to expand the powers of the pharmaceutical, bringing to the attention of physicians, patients, and their families the chemical, social, and economic needs of disordered sleepers. The questions of chemical toleration and need are perverted by the presence of the social, because patients are enticed to balance the side effects of drugs with the social effects of spatiotemporal estrangement.

In October 2005, at a weekly staff meeting, Dr. McCoy took the floor, brandishing a new canister of Jolt cola, about a liter in size and shaped like a battery, with a positive and negative charge indicator on it, as well as a charge rating along the side. It was the latest addition to his collection of hypercaffeinated beverages, prominently displayed in the communal workroom, which all of the physicians used on a daily basis. He went on to explain that Jolt was also packaged in a new green canister that boasted the drink's guarana content instead of caffeine, which McCoy explained was a way the manufacturer could market the drink as "natural." Dr. Richards remarked that adolescent and preadolescent caffeine intake had increased exponentially in recent years, and he reminisced that he was old enough to remember when soda was consumed only on holidays, because it was too expensive to drink on a regular basis. Dr. MacTaggert asked about the effectiveness of drinks containing taurine, which Richards explained was only an amino acid; McCoy opined that taurine was just another thing that manufacturers put into products to appeal to extreme sports enthusiasts and those interested in "natural" stimulants. This vignette was a recurring one in the clinic. Whenever McCoy purchased a new hypercaffeinated drink, he brought it to share with the collected staff at lunch. Also, the growing awareness of the amount of caffeine consumed by patients and nonpatients alike often led to group discussions about this behavior and the impacts that it might have upon sleep. One of the persistent precautions among sleep hygiene practices is to "avoid caffeine within four to six hours of bedtime,"[5] and evidence of contrary uses on the part of individuals inevitably led to discussion. It was not uncommon in the clinic to hear reports of patients who admitted to consuming more than a dozen caffeinated drinks per day. This was corroborated in my conversations with disordered sleepers, who frequently discussed self-medication through overcaffeination, with one individual remarking that she estimated her daily intake of coffee to be about twenty mugs per day, or about five liters. One might rightly expect to be unable to fall asleep with so much caffeine consumed in a day, but for most disordered sleepers, the amount of sleep debt was counterbalanced with constant caffeine consumption. When the caffeine is finally removed as a stimulus, as before bedtime, the desire to sleep is exposed. But caffeine is only one of many contemporary sources of chemical stimulation available to the sleepless, most remarkable in that it is licit while many of its chemical brethren are outlawed.

Four months previous to McCoy's show-and-tell, Dr. Connors, a pulmonologist who spent most of his hospital time in the emergency room and intensive care unit, claimed that the staff were increasingly seeing methamphetamine users in the hospital. MacTaggert asked if they were from the city or from rural areas in the surrounding region, in response to which Connors explained that the three methamphetamine users who were currently in intensive care were all from "the community," in reference to the inner city, MCMC being the emergency room of choice for city services. Methamphetamine users ended up in the hospital, ostensibly for treatment, after their behavior provoked police to stop them. This prompted Connors to ask about the toxicity of methamphetamines as compared with coffee. One of the sleep technicians explained that methamphetamine use in his hometown was so rampant that Sudafed was now treated as a prescription drug, because canny users could distill methamphetamine from Sudafed and other common components. The technician went on to explain that kids at school were selling drugs supplied to them by their parents, which had led to a crackdown on pharmacy sales of cold medications. Connors asked why people use methamphetamines at all, since 40 percent of the population already uses caffeine. This sort of misconception of caffeine as on par with illicit drugs of methamphetamine's potency evidences clinical attitudes toward caffeine use: it is addictive and problematic. The reporting sleep technician explained that methamphetamine provides an instant sense of euphoria, unlike coffee, which has to be consumed cup after cup, indexing the difference between the licit and the illicit: the former category being slow and moderate in effect, the latter being risky because of the speed and intensity of its effect. A visiting doctor explained that Red Bull use is similar to methamphetamine use in that it provides an instant rush, which in turn reminded him of the early 1980s, when people were doing "street speed" (high levels of uncut caffeine—two grams), which was a fad that quickly expired. Dr. Pym happened to have a chart on hand of a methamphetamine user who was going through withdrawal during her sleep study; the chart, Pym said, looked indistinguishable from narcolepsy, with bouts of sleepiness throughout the day and very disrupted sleep. McCoy then told how he had recently been in a skateboard shop and had come across a tee shirt (available in kids' and adults' sizes) that read "feels like the first time" and featured an anime drawing of a vial filled with methamphetamine rocks, in response to which the collected staff expressed dismay.

What is quickly revealed in this conversation is the lack of newness in stimulant use in the United States and the seemingly persistent desire on the part of Americans to be pharmaceutically charged. In light of this, could it be any wonder that Cephalon is pushing for broader applicability of Provigil as a nonaddictive stimulant? Just being awake and alert is sometimes cause enough for the normal sleeper to indulge in chemical stimulation. Disordered sleepers expose how pharmaceutical and chemical stimulants blur in their effects: caffeine promotes wakefulness, while sleep aids induce sleep. Caffeine, modafinil, and amphetamines each act on the body in different ways, producing varied effects, but all promote wakefulness, and what distinguishes them is as much biological as social, as much natural as cultural. Similarly, hypnotics, sedatives, and non-benzodiazepines all produce sleep, albeit in different forms and through different mechanisms. The use of everyday chemicals and medical pharmaceuticals, prescribed or not, is the normal condition of American sleep. Some opt out of this rule, but they are held to the same expectations of sleep and wakefulness, productivity and quiescence, rhythm and order, as those who partake of these chemical forms of life. There is no boundary around the chemical-everyday: contemporary everyday life is as much chemical in its contours as it is social and biological, natural and cultural.

In this context we might rethink the contemporary therapeutic milieu: the lives of disordered sleepers are littered with inopportune eruptions of sleep, accidental but unsurprising sleep, from narcoleptics who nod off when excited to the case of a woman who would fall asleep at parties, be unrousable, and then be taken home by strangers (to their homes, not hers), and find herself waking up on strange couches in strangers' homes. The inappropriately sleepy body disrupts the very conception of appropriateness and everyday life's rhythms; the disordered sleeper disrupts the spatiotemporal order of everyday life. The regular body is the body that lays the foundation for the everyday. The imposition of inevitability—that the everyday must unfold in particular ways and that all bodies should behave similarly—creates parity among actors, in this case between individuals and society, between one and the masses; it renders them isomorphic and subject to the same rules. In the social formulation of the inevitabilities of American capitalism and expectations of human desire, these isomorphic bodies are rendered congruent in rough fashion and inseparable. In order to achieve this level of parity, all actors, individual and institutional, depend on being rendered abstract, on signification over and against the material, on universality versus the particularity of individual lives. These formulations of desire depend on the relation of the masses versus the individual—of the integral and the everyday. Attributing inevitability and diverting it from the abnormalities and insistences that occur in individuals and institutions produce a logic of equivalence among actors; individuals become like institutions, and institutions like individuals. It is in this realm of equivalence that bodies enter the domain demarcated as biopolitics, for it is only through the discursive construction of the natural (the bodily, the biological) and the social (the cultural, the economic) that our desires are pitted against one another and become subsumed in the projects of integral medicine. Our desire for sleep is accepted as biological and natural, despite its intimate complexities.

An overlooked aspect of biopower is the construction of the inevitable over and above its usual insistence on the production of the normal: if biopower is concerned with conceptualizing the potential of bodies, human and otherwise, then essential to any understanding of the body is what the body *must* do, how regular its behaviors should be, how flexible its potentials are, what its desirous and intimate capacities are. If one is to tackle the concept of inevitability from the perspective of its constructed nature, it becomes simply the by-product of larger biopolitical regimes; the category of the inevitable is produced in the increased control of bodies, a spatiotemporal expectation of normalcy that aligns individual bodies with institutional orders. Control regimes depend upon establishing equivalences among discrete objects, on capturing diverse phenomena through governmental means, and on inducing a desire in individuals to align their rhythms with what is taken to be inevitable, to be normal. In this case, the inevitable is the consolidated model of night-time sleep paired with constant vigilance throughout the day.

For each disordered sleeper, normative expectations of the inevitable are put into question, including narcoleptics who spontaneously fall asleep, insomniacs who fail to sleep through the night and insist on naps, irregular sleepers who fail to align their circadian rhythms with dominant spatiotemporal orders. All of them are subject to the designs of control, and some of them strain against the efforts to fully capture them, to move them from the abnormal to the pharmaceutically regularized. As the case of Provigil shows, the diagnosis of new sleep disorders, or rather "waking disorders," indexes a forthcoming expansion of the field in which such models of subjectivity circulate. This, in turn, induces the very category of the inevitable to change. Early in the twenty-first century, Americans are moving steadily away from the possibility of taking a daytime nap and more deeply into the requirement to remain ever vigilant throughout the day, by chemical means if needed. With the intensifying of the spatiotemporal rhythm of everyday life, bodies are becoming more akin to society: bodies must be able to anticipate change, they must desire new intimacies, and constant alertness means that individuals are prepared whenever they are called upon, by work, family life, schooling, or any other social obligations.

In the course of a few short years, excessive sleepiness may successfully become the new erectile dysfunction, the new depression, if Cephalon and its compatriots in marketing pharmaceuticals throughout American society have their way. It will move from the realm of the undiagnosed to the overdetermined. We will no longer have to depend on one ubiquitous drug, sold only through prescription; sleepiness will be relegated to an over-the-counter, direct-to-consumer diagnosis. As a result, we may no longer accommodate our lifestyles to our desires for sleep; we may no longer simply become sleepy; rather, new desires will be bootstrapped into being, new diagnoses will be made available to patients and physicians alike. Rather than creating more stimulating environments or allowing flexibility for daytime sleep, we may take pharmaceuticals, like caffeine before them, in order to maintain vigilance in increasingly quotidian workspaces and in regimes of increasingly taxing demands on time. Stemming from this will be a tacit acceptance of these banal spaces, leading, inevitably, to a spiral into spatiotemporal orderings of everyday life that replace one form of life with another, with intimate obligations controlled by the demands of capitalism. The logic of nature, primordial or modern, is replaced with the logic of the pharmacy and the natures it can produce. The disordered body is more fully controlled through these intimate strategies, producing dependencies that, in turn, help to circulate capital, as coffee, sugar, and tea did before pharmaceuticals. The flexibility of the body and its desire for sleep becomes supplanted by the constructed inevitability of capitalism. In the process, the individual body is replaced by the slumbering masses—populations to be regulated and managed. The necessary politics to contest this biopolitical order realigns the everyday with the manifold capacities of the body, with embracing the multiple desires for sleep, with resisting the imposition of inevitability on the everyday, on work, school, recreation, and family life through the proliferation of desires and intimacies too manifold to be controlled in their entirety.

7

....

Early to Rise
Creating Well-Rested American Workers

> Other individuals have a circadian rhythm that is perpetually out of
> synch with that of the environment, making adjustment difficult, or
> impossible. Individuals with the delayed sleep phase syndrome often
> present with sleep-onset insomnia, lying in bed unable to fall asleep
> until their biological clock permits.
>
> —from the pamphlet titled
> *Practical Guide to Treating Insomnia*

THE PAMPHLET *PRACTICAL GUIDE TO TREATING INSOMNIA*, IN-
tended for distribution to physicians by the National Sleep Foundation (NSF), describes the primary consequence of insomnia this way:
"The direct and indirect costs . . . place a tremendous economic burden
on society and employers." Only after listing a number of costly side effects of insomnia as well as its co-morbidities do the pamphlet's authors
suggest that "quality of life is negatively impacted and family and social
relationships suffer" as a result of the disorder. How is it that the first
concern of a sleep disorder is the economic impact it might have, not for
the disordered sleeper, but for society and employers? Why, in the case
of insomnia, are the needs of the individual minimized in relation to the
sleep disorder's broader social impact? This might have resulted from the
medical view of insomnia as a secondary disorder or from the cultural
acceptance of insomnia as a result of the behavior of individuals; it may
also have resulted from the intensified interests regarding sleep and its

relation to workplace performance and social integration. The National Institutes of Health make a similar claim in the *State-of-the-Science* pamphlet regarding "Manifestations and Management of Chronic Insomnia in Adults": the social effects of insomnia are greater than those experienced by the disordered sleeper. They argue, "The impact of sleep disruption goes beyond the insomniac. . . . Employers of those with insomnia suffer when their work performance is affected." Insomnia is clinically regarded as a complaint rather than a disability qua disability; it is treated only when patients complain about its presence. If a patient is able to live with her or his insomnia, then it remains untreated. But, as both the NIH and the National Sleep Foundation argue, there is a social burden in the side effects of insomnia, ranging from economic losses to diminished work performance. Not only is insomnia socially constructed through recourse to arbitrary spatiotemporal orders, but it is also culturally compounded by the expectations of disordered sleepers and their friends and families and by their social obligations. Added to these complications are the economic desires of employers and governments, who strive to capture as much labor from individual bodies as those bodies allow; insomnia and its side effects surely limit the labor potential of bodies, ensuring that insomnia is the complaint not simply of individuals but of society more generally.

The bifurcation of the moral and the economic involves two faces of the same coin. These complementary discourses, the moral and the economic, serve as mechanisms to align the desires of individuals with the dominant spatiotemporal order of American everyday life. Although they can focus on disparate behaviors and employ quite different language as their means, they are founded in the same concern and focus on the same end. I take seriously the influences of the Protestant ethic in my attempt to elucidate how this concept has insinuated itself into the spatiotemporal formation of American everyday life. This ethic, this moral basis of desire, is based in a profound materiality that depends on a chemistry of pharmaceuticals, social expectations, and economic pressures, which formulate diverse desires and intimacies that bind individuals to institutions and everyday life. I focus first on insomnia and the assumptions that render it dangerous for society. This depends on the unification of many forms of sleep into one pathology, namely, insomnia. I follow this with a discussion of night work, another spatiotemporal order that often leads to disordered sleep. Taken together, these two other forms of desire for sleep, insomnia and shift work, help to evidence the pernicious role of the economy and morality in American thinking about normal and abnormal desire for sleep.

Insomnia has long been a common sleep-related complaint in American society. As the NSF documented in its 2002 Sleep in America poll, 58 percent of Americans identify from one to four of the symptoms of insomnia as occurring in a given week, and 35 percent of respondents claim insomnia symptoms running for an entire year.[1] The symptoms that are nosologically identified as related to insomnia include "difficulty falling asleep, waking a lot during the night, waking up too early and not being able to get back to sleep, and waking up feeling unrefreshed."[2] These symptoms are actually distinct forms of insomnia, the first being sometimes designated as "primary insomnia," the second and third complaints being related to "secondary" and "sleep maintenance insomnia," and the final being related to "fragmentary" sleep architecture, or frequent awakenings associated with a lack of deep sleep. These various disorders fit together under the rubric of insomnia because of the social construction of the umbrella disorder, precisely that any difficulty related to inducing sleep or maintaining sleep quality is termed insomnia. Primary insomnia is among the sleep disorders that are socially constructed and culturally compounded: it exists because of a disparity between when individuals desire to go to sleep and when they can go to sleep.[3] As a medical complaint, insomnia is often seen as an inability of the mind to will the body to sleep, a behavioral problem, not biological one. This interpretation holds despite increased scientific attention to the structure and forces of circadian rhythms.

How is it that a complaint like insomnia can be such a widespread phenomenon in American society? Although the NSF's statistics might be suspect because of a relatively low sample size, they complement the increasing industry of sleep aids in the United States. A "pathological" desire for sleep on the part of Americans may be the result of individual circadian rhythm/bedtime dysfunction; it may also be the result of the dominant spatiotemporal order of American everyday life being at odds with human physiological norms. The need to induce sleep is structured both by the need for consolidated sleep (unless one works at one of the very few businesses that permit workplace napping) and the institutionalized start of the day. According to NSF statistics, the actual pattern of sleep of most American adults who have stable work times is a sliding one: they begin the week sleeping about eight hours, and as the week progresses they sleep less each night, until, on Friday night, they are able to sleep longer into Saturday morning. In this way, Americans are able to catch up on their mounting sleep debt in time to begin the process over again. As a result of the increasing loss of sleep through the week, individuals are able to fall asleep easier and easier throughout the week, but other social obligations remain spatiotemporally constant.

This, however, pertains only to sleep onset insomnia. Sleep maintenance insomnia presents itself quite differently and is prone to have quite different effects on individuals. Sleep maintenance insomniacs wake after sleeping for a few hours—the length of time can vary between one and three sleep cycles, depending on the individual—only to be unable to fall back asleep. Much of sleep maintenance insomnia is not a sleep disorder at all but simply a social disorder; it is problematic to the extent that social obligations are interfered with by sleepiness and attendant stresses. From the perspective of biphasic norms, sleep maintenance insomnia might not be perceived as an abnormality but rather as the result of hundreds of thousands of years of evolution. Moreover, treatment of sleep maintenance insomnia through pharmaceuticals is often ineffective if not dangerous, as evidenced by cases of sleep-related eating disorder caused by the use of sleep aids, especially zolpidem (Ambien). The clinical and scientific model of Homo sapiens' average need for eight hours of *consolidated* sleep may have been largely to blame for the continued understanding of sleep maintenance insomnia as physiologically abnormal rather than as primarily a social problem—for the individual and her or his family, coworkers, and friends.

If, however, the clinical model of human sleep were brought into alignment with what is known about the history of human sleeping patterns, it might be argued that individual patients are not to blame for their inabilities to sleep. Rather, the dominant spatiotemporal ordering of industrial and postindustrial American everyday life, with work and school schedules that privilege larks and those who tend toward consolidated sleep, makes some desires for sleep pathological and others healthy. Consolidated sleep, as it has been accepted, is largely the by-product of the industrial workday, which began as a dawn-to-dusk twelve- to sixteen-hour stretch and shrank to an eight-hour period only at the turn of the twentieth century. Consolidated sleep is the result of fatigue, not evolution, and one's ability to sleep through an eight-hour stretch depends on a certain level of physical exhaustion. Light sleeping might have been selected for evolutionarily, and it may be a great benefit for the majority of those Homo sapiens individuals who are subject to predation or who tend to livestock and necessarily maintain fire throughout the night. Deep sleep, in this evolutionary assumption, is dangerous, if not life threatening, and only with the invention of protected homes have sleeping behaviors become protected from predators. Also, as shown in the work of James McKenna, light sleeping has a benefit for childcare in cosleeping situations, and humans are prone to frequent partial arousals throughout sleep, although they go largely unnoticed by the sleeper. As a result of the combination of lifestyles that promote chronic sleep deprivation and social time–spaces that produce noxious noise and other disturbances, evolution has been turned on its head, and light sleeping has become a detriment, while those who sleep deeply find themselves comparatively well rested.

Early in the twenty-first century, insomnia's co-morbidities are being attended to by physicians and researchers. This is the result of the increasing view of insomnia not as a primary medical complaint but as a symptom of other health-related issues. This view results in using medications "off label" or attempting to treat what is perceived by the patient as a secondary complaint rather than treating the insomnia itself. For example, although feelings of depression may lead individuals to spend more time in bed, contemporary research increasingly attests to a greater prevalence of insomnia complaints among the depressed. As noted in the NIH's *State-of-the-Science* pamphlet on sleep medicine, the most popular drug prescribed for insomnia is not one of the benzodiazepine receptor agonists that have been developed for the cessation of insomnia, but trazodone, a drug primarily used as an antidepressant. Most insomnia responds positively to sleep aid pharmaceuticals—antidepressants and benzodiazepines as well as antihistamines—but most patients with chronic insomnia develop tolerances to the drugs they take, which often requires them to cycle through prescriptions in an attempt to retain the chemical gains of the drugs. Other patients, however, never respond positively to pharmaceutical treatment and seek other means to curtail the negative effects of being out of synch with the dominant spatiotemporal ordering of everyday life. This can lead some to invest in another mode of desire, a different intimacy with sleep, labor, and the spatiotemporal order of everyday life.

Night work has been a constant sector of the workforce since at least the Middle Ages, although this history is often obscured to render the so-called twenty-four-hour society newer than it is. Not all work can be carried out through the night, but as Jacques Le Goff notes, at a time when most work was done in the home, it was easy and profitable to work into the night;[4] similarly, because of the cost of lighting, it was easier to provide light for a room of workers than to leave each worker to his or her own devices in his or her own home. This rationale carried through the industrial period and persisted, although in modifying forms, throughout the twentieth century. Exclusive night workers, however, may be a relatively new breed or one that arose during the industrial period through the use of electric lighting. Night work, prior to the industrial period, was marked by its flexibility: a person might work one night and spend the following morning in bed before working through the afternoon and evening. In societies like the United States, however, night work and shift work are often inflexible for those who work such hours. Night shifts are relegated to junior members of an employer's staff, who, upon weathering the adverse effects of night work and being desynchronized with dominant social life, may finally attain a normal work schedule, which is the case in professions such as law enforcement, medicine, and emergency services. A host of health detriments are associated with night and shift work, including obesity, depression, and gastrointestinal and pulmonary distress, and so are a number of social disorders associated with working schedules at odds with one's family and friends. About 10 percent of shift and night workers exhibit symptoms associated with shift work sleep disorder (SWSD), which presents itself as persistent insomnia, because one's time for sleep is invariably at odds with one's circadian rhythms.[5] Individuals with SWSD symptoms often find no relief in pharmaceutical treatments and struggle, like most other disordered sleepers, to align their expectations of desirable sleep with the spatiotemporal demands of society. Only in the case of SWSD has sleep medicine begun to confront its own presumptions about the need for consolidated sleep, because one of the few effective treatments for those who exhibit SWSD is regulated biphasic sleep.

American sleep scientists hold that the human norm of circadian rhythms is based on a twenty-four-hour period, with minor variations (early models held that the period was twenty-five to twenty-six hours in length), and these rhythms constitute two primary functions, Process S and Process C. These terms respectively refer to the homeostatic sleep drive and the circadian drive for wakefulness. The models that circulate in sleep science are all based on consolidated sleep patterns; for example, the sleepers whose sleep generated the models slept for eight hours each night, without a nap throughout the rest of the day. Lacking are similar models based upon biphasic sleeping behaviors. In the models depicted in Figures 2 and 3, the arrows at the top of the graphs indicate the urge to sleep, with sleep being most insistent around midnight. In Figure 2, notice that the desire for sleep is at its lowest around 7 A.M. It then increases throughout the day, until, at midnight, the idealized normal sleeper goes to sleep. In Figure 3, the midday and after-work dips in wakefulness represented by the waveform indicate "sleep's propensity." Although these models purport to evidence the same internal drives, Figure 3 remarkably differs from Figure 2 in its attention to "clock-dependent alerting" (the equivalent of Process C in Figure 3). In both models, the rebound from midday sleepiness is dependent upon internal alerting systems, in this case based upon circadian cues that emanate from the liver, itself cued to food intake.[6] Contrary to popular conceptions, lunch wakes us up in the afternoon and slightly masks our building desire for sleep. Napping, however, might produce a similar rebound, but it would dramatically alter the representation of Process S. If napping were accounted for, the desire for sleep would appear as a more evenly distributed force rather than one that depends upon widely disparate peaks and nadirs.

Many sleep patterns, such as insomnia and SWSD, are experienced as disorders, because they prevent individuals from integrating into what is understood as normative everyday life. This is particularly true in relation to work hours, but, since work hours provide a foundation from which to articulate other spatiotemporal activities, this obstruction extends throughout everyday life. Insomnia and SWSD are pathologized forms of a spatiotemporal alienation that late risers, or owls, often confront, unlike their counterparts, larks, who face little difficulty fitting into the traditional American workplace or school. Owls often experience insomnia in that they fail to be able to sleep when they should in

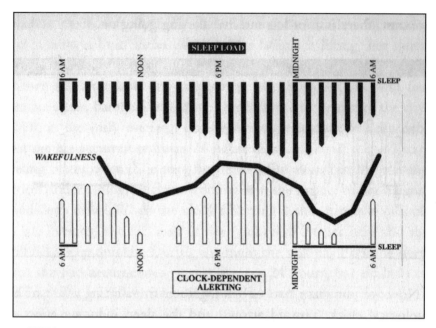

FIGURE 2. William Dement's representation of the circadian rhythm in humans. The "sleep load" is counterbalanced by "clock-dependent alerting," which promotes wakefulness and indexes sleepiness. William C. Dement and Christopher Vaughn, *The Promise of Sleep* (New York: Delacorte Press, 1999), 84.

order to work a 9 to 5 day. Even if owls were to take work whose timing coordinates with and is thus more amenable to their desire for sleep, this arrangement would skew their alignment with spatiotemporal regimes outside the workplace, replacing one disorder with another. While sleep disorders are accepted as being treatable through medical interventions, social disorders often cannot be treated in the same way and demand that the disordered sleepers alter social obligations to meet their desire for sleep. As might be expected, the greatest resistance to the cure for these ailments is the individual's understandings of cultural normativity and the expectations of what it means to be a part of American everyday life.

At the end of my first year of fieldwork at MSDC, a senior researcher, Dr. Xavier, presented the case of a thirty-five-year-old man, Frank, who, by his own admission, had always preferred night work. He was able to sleep between six and seven hours during the day, managing, in Xavier's

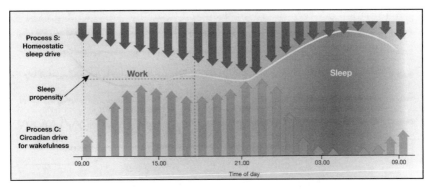

FIGURE 3. A graph of the circadian rhythm from a pharmaceutical guide for patients and clinicians. The day is demarcated into "work" and "sleep" with very little time left for other activities. Pharmaceuticals, in this representation, are expressly for the production of laboring bodies. Provigil (modafinil) product monograph, Cephalon, Inc., November 2004, Frazer, PA, 7.

words, to "have a life" in what remained of the day. Frank had been diagnosed with diabetes three to four years before seeking medical help for his disordered sleep; what finally led him to seek this medical help was an increase in his daily sleep requirement to upward of twelve hours per day. Xavier remarked that Frank's life was "a shambles" and explained that he also had signs of depression. Frank had not had a date in the previous fourteen years, which may have been largely the result of his choice to work nights, but the situation had recently been compounded by his escalating health concerns. Frank was reported as often having thought about taking his own life with a gun, which drew some exclamation from the rest of the staff. His hope in seeking medical help from the staff at the sleep clinic was to reduce his time-consuming sleep needs. After the presentation of the case, some discussion ensued about advising Frank to alter his work schedule to a daytime one. This conversation foundered for lack of knowing whether Frank was amenable to such a shift in his social use of time. He was being shifted as a patient of Xavier, a staff neurologist, to a patient of Dr. Samson, one of the staff psychiatrists, because of his suicidal ideation and symptoms of depression. Frank's "preference" for night work, his desire for it and daytime sleep, disallowed romance, although chronic depression also seems to have played a role. Frank's case shows how dominant spatiotemporal regimes of work make demands on individuals and how this necessary coordination between desire for sleep and desire for work is integral to the production of individual subjectivity. This is the case in one's life as a worker but also in one's participation in American everyday life more generally. When such coordination between individuals and institutions fails, compounding consequences often occur.

Like Frank, many disordered sleepers often experience a feeling of being unable to inhabit what they understand as the normative every-day life that their friends and family experience, which potentially extends to all of social life, in workplaces and schools and in daily activities elsewhere. Normative models of desire for sleep and everyday institutions coproduce each other with ideas about what times of sleep are most cultur-ally appropriate, efficient, and natural, structuring ideas about the most culturally appropriate, efficient, and natural workday, as well as ideas about transit to and from work and what social activities might precede and succeed work. These models, in turn, are relied upon in order to diagnose abnormal sleep, or sleep that is rendered as inappropriate, in-efficient, and unnatural; prime among these pathologized patterns of sleep is SWSD. Individuals who exhibit symptoms associated with SWSD are affected by social demands in intensified form, with their aberrant work schedules leading to further difficulties, which, in turn might lead them to make work-related decisions that exacerbate social estrangement as well as continued medical concerns. This was evident in Frank's case, in which the desire for late sleeping led him to take a night job, and his work schedule then precluded him from some aspects of social life. This es-trangement reinforced his decision to work nights. Eventually, this night-time working schedule led to deeper medical problems, including diabe-tes and depression, which led to further social estrangement, including an inability to continue working his previous shift as well as intensified depression. Reintegrating into day-based everyday life may have produced even more complicated problems for Frank, because his inability to meet the demands of the everyday chafed against the normative model of sleep, as is concretized in the model of sleep that accepts consolidated sleep as its foundation. Models, in this case, produce pathologies, which, given a model based upon biphasic sleep, might not be pathological at all; models, in this case, also produce social disorders, as those who are relegated to shift and night work invariably confront social estrangement. Altering the demands of American institutions might concomitantly alter the model of everyday life, but the promise of new therapies leads to the pursuit of other means of rethinking work during undesirable hours.

Individuals with disorderly circadian rhythms serve as a caricature of modern American sleepers, held as they are in a pattern of sleep and social life that is tenuously balanced upon an intimate network of cultural expectations, social and economic obligations, biological predispositions, technological prostheses, and chemical stimulants. Most Americans are able to negotiate such demands, but these negotiated lives are the result of years of enculturation, which begins in childhood and intensifies during schooling.

Take, for example, Timothy, a white, middle-aged man, who at the time of our interview had evaded diagnosis and confronted feelings of intense sleepiness throughout his life. This was coupled with an inability to meet the everyday spatiotemporal demands of American everyday life, stretching back to his youth:

> I remember in seventh grade not being able to stay awake in school. Right up until sixth grade I remember getting good grades for the most part—all As and Bs. I do remember in seventh grade I changed schools, and there was one class in particular—it was math class—I began to uncontrollably fall asleep in class. I just thought it was the subject; I thought it was the teacher, because he was *very* boring and monotone and I didn't think anything of it. But then from then on in every grade and then in high school it was almost every single class that I was falling asleep. In high school I was a big coffee drinker in the morning. I actually used to take No Doz—they're just caffeine pills, I believe. In high school I started popping No Doz and drinking a couple cups of coffee before I left for school.

What Timothy brings into relief is a narrative shared with many sleep patients, namely, that of his dependence on self-medication through caffeine (whether coffee, soda, or in pill form) that students resort to in order to normalize themselves. This regime of normalcy—a mix of chemicals and social obligations meted out through the body, producing a common form of desire and intimacy—is the model upon which all later, adult iterations found themselves. This is the case in the context of work and family life and whenever else sleep and the everyday overlap. This complex materiality, bringing together kinds and scales of quite different orders (individuals and institutions, chemicals and consumption, behaviors and natures), defies the reductions that are often accepted as the basis of the economy, culture, and society; these intimate spatiotemporal orders are the natural state of American everyday life and rely upon the production of normative ideals that necessitate interventions that produce desire for sleep that is orderly and normal.

8

· · · · · · · · · · · · · · · · ·

Chemical Consciousness

MY FIRST INTRODUCTION TO THE COMPLEXITIES OF THE DE-sire for sleep and its various intimacies came early in my fieldwork at the MSDC. In my third month of fieldwork, Dr. MacTaggert, one of the staff pediatricians, presented the case of a young girl who experienced severely disordered sleep, the result of which was a strained family life for her entire immediate family, which affected her parents' work lives. MacTaggert presented the case as that of a "fascinating girl" and passed around her sleep chart, remarking that her "recipe was on back," in reference to the array and dosages of drugs she was taking. Her sleep chart evidenced that the girl seemed never to sleep for more than a two-hour block and generally only in hour-long blocks throughout the day and night, with more wakefulness during daylight hours, but no firmly consolidated sleep through the night. Prior to seeking consultation for her disrupted sleep, she had been diagnosed with bipolar disorder with schizophrenic tendencies, and MacTaggert needed to take this prior condition into account in diagnosing and treating the girl's aberrant sleep. The girl's sleep disruptions coincided with depressive phases, which is not uncommon for depressives, who often show high rates of insomnia, although they may spend proportionally more time in bed or "resting."[1] When the girl did awake at night, she lay in bed in the dark, not getting up to do anything, including eat, which required her parents to bring her food. She had already been given a prescription sleep aid, Sonata, at bedtime, but MacTaggert wanted to replace this with a longer-acting

171

drug, because Sonata is generally effective in curtailing only sleep-onset insomnia. With the girl's repeated sleep maintenance complaints, she required something that would continue to induce sleep throughout the night. In MacTaggert's words, the girl was "very, very disordered," adding, "The prognosis is not good." MacTaggert then listed the five drugs that the girl was already on, and Dr. Richards, the center's director and head neurologist, reacted with an exclamatory "God," dismayed with the number and dosage of drugs as well as the particular cocktail of antipsychotics, antidepressants, and sleep aids. Dr. Pym, one of the other staff pediatricians, then suggested that the clinic produce an actigraphy of the girl's sleep, but Richards discouraged doing that because it would not have any immediate therapeutic effect, and the clinic already had evidence enough of the girl's chaotic sleep patterns. Pym was hopeful that because so little information had been collected on such conditions of erratic sleep that an actigraphy might prove useful in the future, and he opined in closing the discussion of the case that there was "not a lot to do with kids on so many psychotropic agents."

At another staff meeting in my second year of fieldwork, Dr. McCoy, the codirector of the sleep center and one of the head pulmonologists, announced that a recent finding from the Mayo Clinic, in southern Minnesota, posited a causal link between Mirapex use (used primarily for Parkinson's disease, but also sometimes for certain sleep disorders, such as REM behavior disorder) and compulsive gambling. McCoy wanted to bring these findings to the staff's attention because the national interest in the tentative statement made by the Mayo Clinic researchers was sure to affect life in the clinic. This was corroborated by one of the clinic's administrative assistants, who reported that three patients had already called the clinic to confess their gambling problems. McCoy noted that it was an uncontrolled study and that he was bothered that the national press had so vigorously picked it up, attributing this attention to its origin in the Mayo Clinic, generally regarded as having an uncontested prestige. Dr. Samson, a staff psychiatrist, noted that Mirapex had been known to cause compulsive sexual behavior and overeating, so gambling made sense, but they had yet to witness any such cases in the clinic, and no patients had previously admitted any gambling problems. It was only with the Mayo Clinic's announcement that patients seemed moved to link their gambling habits with their Mirapex use. Dr. Banner, a neurology resident, explained that when they prescribed dopaminergic agents, they asked patients about mania tendencies, but he had yet to hear about gambling from any patients. One of the sleep technicians complained that such a study might act to facilitate false claims against the pharmaceutical company by people who already had gambling problems but now attributed them to Mirapex; this met with general agreement among the assembled staff. McCoy explained that since it was an anecdotal report, it should not be taken too seriously by the staff, and they should do what they could to curtail the association of gambling with the drug's use. But one of the clinic administrators emphasized that since it had come out of the Mayo Clinic, many people would accept its validity despite its uncontrolled, anecdotal nature, acknowledging the prestige of the source rather than the lack of scientific rigor.

Although the physicians were primarily concerned with the health of patients in both of these cases—the young girl who exhibited erratic sleep and the relationship between Mirapex and compulsive gambling—what intruded upon their ability to diagnose the proper treatment potential of drugs was the possible negative impacts such treatments might have upon the social relations of the disordered sleepers. In the case of the young girl, her previous diagnosis of mental instability (namely, bipolar disorder and schizophrenia) complicated the effect of an additional pharmaceutical intervention on her sleeping behaviors. The problems that her family faced because of her unpredictable sleep were what necessitated treatment of her sleep disorders, however they possibly could be. Although Pym's closing statement, that there was "not a lot to do" with such patients, might be read as a defeatist response to the situation, this was offset by MacTaggert's willingness to pursue treatment. This meager optimism might be perceived as an acceptance of the possible curative abilities of the girl's intimate network; Pym and MacTaggert, as practiced pediatricians, repeatedly evidenced a faith in the influence of parents and the dedication of young patients in ensuring their improvement. The case of the relationship between Mirapex and compulsive gambling is a bit more complex, in its desires, intimacies, and the forms of life it produced.

Because of its prestige, the Mayo Clinic often confounded the daily practice of the clinicians at the MSDC during my research there. This was redoubled because of the Mayo Clinic's proximity to the MSDC and the easy relationship between the local National Public Radio station and the medical research industry throughout Minnesota. For example, researchers from both the University of Minnesota's medical school and the Mayo Clinic were often interviewed on local National Public Radio stations, and the director of the MSDC made at least seasonal appearances on a morning call-in show that local Minnesotans telephone in search of medical recommendations for their sleep complaints. To complicate things further, the Mayo Clinic has acted primarily as a research institution, often referring patients to the MSDC; although this has been necessary, it was often remarked upon as a minor source of frustration, because the clinicians would then have to intervene in the relationships between patients and researchers at the Mayo Clinic. Finally, the MSDC played an important role in the definition of REM behavior disorder (RBD) (although a formal nosological definition did not appear until 1986),[2] and the more recent research associating RBD with Parkinson's was conducted solely at the Mayo Clinic, albeit dependent on brains and patient records that were donated from patients seen by physicians at the MSDC. McCoy's announcement of the Mayo Clinic report regarding Mirapex should be understood within this context, because many RBD patients have been seen at the MSDC, and the physicians there are perceived within the broader sleep medicine community as some of the foremost experts in RBD and related parasomnias.

How then were these physicians to intervene in what may have been a serious social side effect of a drug necessary to control a sleep disorder? Fixing the gambling problems that may have surfaced in some Parkinson's and RBD patients taking Mirapex would have required removing them from a drug that was otherwise beneficial. Yet the real monetary cost of staying on Mirapex and developing dangerous gambling habits may have resulted in serious financial difficulties for the patients and their families. Treatment possibilities are quite narrow for Parkinson's, with very few pharmacological alternatives, and if the affected patients were on Mirapex for sleep-related complaints, then they had already evidenced intolerance for frontline sleep medications. Thinking through this situation from the perspective of the many desires and intimacies at play brings into relief the production of doubts within the Mayo Clinic's report, doubts that could poke holes in medical thinking through which the clinicians might have potentially intervened. The veracity of the report was what the clinicians critiqued first, but this failed to curtail the complaints of patients who had already been exposed to the report. Instead, what had to be employed was a diffusion of the social side effects of the drug: patients needed to reinvest themselves socially, and the enticement to consider their behaviors reflexively assumed clinical importance. Clinicians needed to probe a patient's social behaviors to the extent that they might understand a patient's intimacies; in order to ensure that the side effects of the drugs were real and not matters of conjecture on the part of patients or other medical authorities, clinicians had to alert patients to the possibility of the drug's potential for new, destructive, albeit temporarily pleasurable, side effects.

Sleeping at inappropriate times is often interpreted as a behavioral problem rather than a biological one, and disordered sleepers bear the brunt of this view; the institutions they interact with may be arranged to ensure their continued sleepiness and failure, despite their best efforts, including self-medication. In the following interview excerpt, Andi recalls how her sleepiness in school was viewed by those around her as her "wanting attention"—clearly related to behavior rather than biology. In her youth she had been diagnosed with infectious mononucleosis, which, in hindsight, she perceived as having interfered with an earlier diagnosis of her sleep disorder, namely, narcolepsy with mild cataplexy.

> I missed a lot of school, and I slept through a lot of school. It's kind of funny—I would need a nap right then and there, and I was out of commission. A lot of it they said had to do with wanting attention. . . . My mother got me up when I was younger, and I was always able to get to school, as a kid. Also, through college, I would try and avoid morning classes because I had trouble getting to them. I could short myself through the week and catch up on the weekend, but I'm not that young and flexible anymore.

Eventually, Andi took it upon herself to arrange her school days in such a way as to complement her desire to sleep. Andi relied upon what she perceived as a bodily flexibility, an ability to short herself through the week as a response to the institutional demands of school, which would inevitably need to be repaid over the weekend. That American institutions have come to rely upon a perceived flexibility in human capacity has only reinforced the assumption that an inability to remain awake through class or work or recreational times is a behavioral decision of individuals, who might then be treated as unruly subjects.

Embedded within a social formation that depends upon strict spatio-temporal stabilities and expectations of bodies, disordered sleepers, diagnosed or not, are forced to find means to support themselves through the course of the day. For those diagnosed with excessive daytime sleepiness, recourse might be made to the drugs formerly reserved for narcolepsy; for those with undiagnosed sleepiness, other stimulants and means are used in order to achieve some sort of everyday normalcy. And for those unlucky few who are woefully misunderstood or misdiagnosed, recourse might be made to exercise equipment, shaming techniques, and other odd control techniques that align individuals with normative expectations. Because of the disciplinary aspect of schooling institutions in the United States, the interpretation of sleep disorder symptoms as behavioral rather than biological problems is more readily accepted, leading to classroom punishment rather than medical intervention. This misinterpretation of the inciting problem can in turn generate further social problems, creating further distance from an accurate recognition of the underlying cause of the inciting problem. This is exacerbated by the widespread cultural assumption in the United States that although we cannot ultimately resist sleep, individuals are able to control when and where they sleep. The acceptance of this model empowers individuals but obscures individual desires, relying instead upon a foundational understanding of all human sleep as necessarily the same; for those who sleep through class or work or arrive at school tardy, these behaviors can be perceived as behavioral decisions, choices to be unruly, and not the fault of the institution. Even where school start times have been recognized as possibly not amenable to many students, those with sleep problems might still be construed as having bad behavior, as not accepting the form of life necessitated by the split between biological and social demands.

This acceptance of sleepy everyday behavior as somehow the fault of the individuals, as if they unconsciously will bad behavior as a reaction to situations that fail to be engaging, is further evidenced by other disordered sleepers. In the following, Kat, a narcoleptic and young mother, made recourse to her bad behavior to explain her "issues in school" but recognized that her classroom performance may have had more to do with her desire to sleep than with her capacity to be attentive in school:

Was I just lazy? Or not paying attention? Or was it because of that [her sleepiness]? I really don't know. I had issues in school, and I guess it was around the time that it started that I was always a good student and did really well and was in the top of my class, and then in high school it just—I found it really hard to concentrate. . . . I did fine in classes, but if the grade was based on homework, it just wasn't happening. I would also sit in class and zone out and not remember most of the class. Looking back on it now I realize that it [the beginning of narcolepsy symptoms] probably was around the same time that my grades started to suffer, but I guess at the time I didn't realize. I just thought I was not into it, y'know?

The internalization of bad behavior is the effect of American forms of desire that rely upon strongly held normative ideals. When individuals are unable to align themselves with these normative ideals, despite their desire and effort, they come to conceive of themselves as abnormal, as disordered. This results in a sense of individual difference: the disordered are those whose difference always hinders their ability to meet normative expectations, but rather than accept physiological conditions as the source of these problems, they see social behaviors as the cause of disruption and the target of control. Those students who remain undiagnosed throughout their schooling often make recourse to means that help them achieve some semblance of normalcy and ease them through their daily obligations. For most of these students, reliance upon unregulated stimulants (caffeine in its many forms) is a given in their self-treatment, which further obscures the physiological underpinnings of their problems and leads to the continuation of those problems.

For many disordered sleepers, especially through their youth when they experience undiagnosed problems, or problems that might be readily accepted as behavioral rather than biological in their foundations, self-medication becomes a necessity. And with the variety of substances through which caffeine can be consumed—which are not limited to coffee, tea, soda and other soft drinks, chocolate, pills, and gum and only seem to expand annually—most individuals are able to find means to insinuate caffeine into their daily routine. Rodney, a white male who was thirty-five at the time of his interview and only recently diagnosed with narcolepsy and a circadian rhythm problem, recounted that he had experienced severe sleepiness throughout his life, but throughout his high school years he had relied heavily on caffeine to be able to maintain himself throughout the day. In his case, the caffeine was in the form of energy drinks and soda:

I was basically self-medicating. I was very religious about time to bed [and] time to get up. It was just my nature. I drank a lot of caffeine—a lot.

How many cups of coffee per day, would you say?

I didn't drink coffee, but I would easily drink, let's say, twelve Diet Cokes a day, sometimes more. Energy drinks—Red Bull, stuff like that. Now [since diagnosis and treatment] I drink two cups of coffee a day, in the morning.

Annette, a young black woman who had experienced excessive sleepiness since the age of five, similarly treated her symptoms with legal stimulants. At the time of her interview, she worked for the U.S. military as a medical administrator and had become increasingly aware of the role that her sleepiness played in her life throughout the course of boot camp, which she recalls in the following exchange:

Were you self-medicating with caffeine?

Oh, god, yes. What is it, ephedrine pills? Me and those were best friends. I always had some. I went through about eight or nine Mountain Dews a day. Um, yeah, and it still didn't help, that's what I found hilarious—that I could go to sleep drinking Mountain Dew like it was nothing. Caffeine pills—No Doz is what got me through boot camp. I always had the shakes, but I thought it was normal. I didn't enjoy the ephedrine, but I took it when I had to.

You said you've had symptoms all your life; how do you think they affected your schooling?

I slept during my classes; I always have. I'm lucky enough that I'm fairly intelligent and functioned [well]. There were parent–teacher conferences a lot during high school, because I had teachers complaining about me sleeping in class, and my mom would just make a deal with them, like "If she drops below an A, tell her to wake up. Otherwise just let her sleep." I slept my way through school; I don't remember much of high school. Every class was nap time. You learn the methods behind it, like sleeping with your eyes open, or ways to position your book [to sleep behind it]—I knew it all.

Annette and her mother both realized that her behavior in the classroom was dictated in part by her desire to sleep; her mother was willing to intervene on her behalf to ensure that she was able to meet institutional expectations of being in school while still abiding by her desires. Annette's later reliance upon caffeine throughout her adult life, while possibly startling, is not exceptional. As it did with Rodney, the spatiotemporal order of American everyday life produces individuals who can function only through reliance upon chemical means, a decision that they are compelled to make for themselves and that is understood as an individual choice rather than a social necessity.

This situation of chronically sleep-deprived individuals who must rely upon chemical stimulants is one that has been produced by expectations of socially appropriate forms of everyday life, including chemical dependencies. However, despite the evidence that the conception of the school and work day might be a central cause in this production of a pervasive sleep debt, we often attempt to place the blame elsewhere, such as on the behavioral decisions of the disorderly, who make easy victims to accuse for their culpability in their sleep deprivation. Embedded in both William Camden's "the early bird catches the worm" and Benjamin Franklin's "early to bed and early to rise, makes a man healthy, wealthy, and wise" are assumptions not only about proper behavior, particularly proper American behavior, but also about normative ideals. What Franklin and Camden presume is that controlling sleep is a behavioral issue and one that, through sufficient effort, "a man" might successfully master his economic, intellectual, and bodily benefit. As many sleep disorders evidence, no amount of willpower can overcome the inevitable demands of sleepiness. But institutions might be ordered differently, producing alternative models of desire for sleep and other intimacies.

9

·················

Sleeping on the Job
From Siestas to Workplace Naps

ROM THE 1970S THROUGH THE 1990S, PREDICTIONS OF THE
way capitalism would "colonize" night were prolific, in both popular culture and academic discourse. What happened in the course of
the 1990s was an inching away from this colonization of night: rather
than U.S. businesses staying open later, service industry jobs were sent
overseas, thereby preserving a diurnal, consolidated sleeping scheme for
many Americans—and for those who serve them elsewhere, although
sometimes at odd hours. For example, Indian call center workers work
through their nights to be awake during our days. It may be argued that
this creeping away from night, this retreat into diurnal practices, is the
point at which the biology of the human body has resisted the efforts of
global capitalism. The resistance to the penetration of night, however, is
an effort to promote specific global cultural norms: it is about the globalization of American expectations of normative sleep. Moreover, this
form of spatiotemporal imperialism privileges specific societies and cultures to the detriment of others: India becomes synchronized with the
United States, Spain with England, France, and Germany; efforts toward
modernity are inescapably tied up in this spatiotemporal order, which is
often posited as a morally superior, natural use of time and space. In this
chapter, I focus on four interrelated events: the disappearance of the siesta
in contemporary Spain, the rise of workplace napping, the retreat from
nighttime business in major U.S. cities, and the synchronization of Indian

call centers with U.S.-based corporations. At first glance, these may seem quite separate phenomena; however, they are all symptoms of the subtle influence of American normative ideals of sleep, an influence that has led to a form of spatiotemporal, rather than strictly spatial, imperialism. This spatiotemporal imperialism takes labor, fitness, and desire as its primary foci, employing discourses of health as its means of insinuation.

Many discussions of economic transformations in the last half of the twentieth century highlighted what has become accepted widely as "time–space compression."[1] This is the idea that temporal and spatial access among (until recently) widely divorced spaces has increased, emblematized in intercontinental flights, e-mail, satellite radio transmissions, and cell phones. This compression is posited as both the appearance and the experience of proximity. In the following I follow an array of formations that have been exported from the United States and western Europe and that, rather than compressing time and space, synchronize temporality over diverse spaces, thereby coordinating diverse bodies and distant societies. Rather than accepting other spatiotemporal orders, the United States, especially, and the European Union syncopate the rhythms of other societies to produce global temporal intensities and rests, metonymic coordinations of workers and work. The effect of this is a sleepy American capitalism exported to the world, a form of capitalism imbued with the familiar spatiotemporal agency of the human body, with periodic remissions and vigilance and a nature of its own, all of which are relied upon by diverse actors to posit American capitalism as natural. These kinds of coordinations depend on aligning local experiences of space and time with global forces; they involve producing everyday orders that make economic and biological sense and are able to collapse cultural expectations into bodily realities, instituting forms of life predicated on particular desires and intimacies.

One model of social arrangement that meets the biphasic model of sleep is that of the siesta culture, in which, around midday, work breaks for mealtime and rest. This allows owls and disordered sleepers to catch up on their sleep and gives larks a midday respite. However, around the world, as local economies are coordinated with more powerful neighbors, siestas disappear. The exemplary case is Spain, which altered the workday to eliminate the possibility of midday naps. Beginning in January 2006, government offices and services adopted a 9 to 5 workday model, eradicating the traditional siesta break in the late afternoon. The hope, of both government quarters and independent lobbyists, such as the Fundación Independiente (a conservative think tank), was that this would force other businesses to synchronize themselves to the 9 to 5 schedule. For example, as published in a Fundación Independiente white paper:

> Today, while Spain's neighbors to the north have already digested their meal, the lunch hour in Spain is just beginning, and often it lasts not just an hour but hours. Such a lengthy meal makes it difficult for employees and managers in Spain and other European countries to make arrangements between midday and 4 P.M.[2]

"Not just an hour but hours." Embedded in this repudiation of the Spanish status quo is an expectation of normal human behavior—that an hour for lunch is surely enough—and that a variance from that norm is problematic. At the local level, these attempts toward spatiotemporal hegemony rely on discursive tactics of shaming, which involve pointing out Spaniards' lack of productivity compared with that of the rest of the European Union, with some estimates placing them 40 percent behind their northern peers,[3] and a deployment of normative ideals: that human beings are biologically suited for a 9 to 5 workday, which has been embraced in nearby France, Germany, and England and has been accepted as a dominant model of healthy sleep. However, as some U.S. businesses have discovered in the early twenty-first century, a midday nap could greatly benefit workers' health—and their productivity. Workplace naps also keep workers at their workplaces for longer hours, since they feel little fatigue at the end of the workday so are more inclined to continue to work.[4] Spaniards may soon find their cultural tradition of the siesta repackaged as the workplace nap, along with a pillow especially designed for desktops.

One attempt to rectify some of the sleep deprivation endured by Americans in the late twentieth century was the insinuation into workplaces of both more attention to the powers of napping and spaces for potential nappers to inhabit. This initiative was spearheaded by Alertness Solutions and the Napping Company, two consultancy firms that, for a fee, would visit a workplace to consult with employers and their employees on how napping might benefit both workers and the economic goals of the company; they also provided advice on how, where, and when to nap. To overcome some of the derision commonly targeted at nappers and napping behaviors, both consulting firms coded their advice in the language of American capitalism; they provided extensive data on the economic benefits of napping for employers. So, while nappers might have felt as if they were the ones benefiting from the change in workplace policies regarding napping, employers were actually the ones who reaped the greatest benefit. The development of workplace napping was an extension of the flexible workday, an attempt lauded in the 1990s as a more humane way of arranging the workday, although this humane workday was not borne out by flextime as actually practiced.[5] Moreover, in order to work flexible schedules, workers were often expected to make other workplace sacrifices, sometimes moving from full to part-time employment, increasing the length of workdays, or accepting pay cuts.[6] Finally, access to flexible work schedules was stratified on the basis of the kind of employment, with managers and those who work nonindustrial jobs having access to flexible work schedules and low-wage and traditional blue-collar employees having limited or no access. Napping in the workplace, conversely, was an activity available to everyone, and the possibility of insinuating it as a workplace option relied only upon overcoming cultural expectations of sleep and its proper times and places. To achieve this end, workplace consultancies were developed to spread the gospel of sleep.

The premise on which Alertness Solutions operates is that modern life has become inherently more time consuming and that a break from nature and natural rhythms of life has occurred through reliance upon artificial social mechanisms, such as work time. Technology and globalization are perceived as evils that have irrevocably altered not only everyday life but also human relationships with the natural world. Thus, for Mark Rosekind, the director of Alertness Solutions, napping is a means to curb an incessant form of life:

> Today, 24-h operations are necessary to meet the demands of society and the requirements of an industrialized global economy. These around-the-clock demands pose unique physiological challenges for the humans who remain central to safe and productive operations. Optimal alertness and performance are critical factors that are increasingly challenged by unusual, extended, or changing work/rest schedules. Technological advancements and automated systems can exacerbate the challenges faced by the human operator in these environments.[7]

Rosekind and his colleagues argue that workplace demands have altered over time, increasingly encroaching on an individual's ability to attain suitable amounts of rest. Although they focus on institutional impediments to adequate sleep, ultimately they charge individuals with the need to advocate for and assert their own desire for sleep. In an interview for the World Economic Forum, reprinted in the publicity packet disseminated by Alertness Solutions, Rosekind elaborates this program in more colloquial language: "Can you think of anyone in our society who isn't working 24 hours a day, has a crazy schedule, crosses time zones, has to get up early for work or has to work through the night?" Work times, however, are distributed across class lines, and the ability to nap—or the workplace perk of napping—is also differentially distributed, with employers who privilege intellectual over physical labor more likely to adopt policies amenable to the sleepy worker.

The Napping Company, led by William and Camille Anthony, takes a tactic similar to that of Alertness Solutions, but it stresses the need to translate scientific and medical knowledge into the language of economics, a language that the company assumes corporate administrators will find compelling. The Anthonys offer the following as examples for altering the institutional rationales that prohibit workplace napping: "Napping at the workplace can improve worker productivity," or, "napping is the no cost, no sweat way to improve worker performance," or "napping can be good for you and good for your company."[8] They suggest that workplaces adopt more ambivalent policies regarding napping, policies that would not endorse napping or mandate it for all employees but rather would act passively to allow workers to nap if they so choose. But, as the Anthonys hint in the following, workplace napping is more complex than simply ensuring that employees will not be dismissed from their jobs for taking a nap:

> A napping policy would simply be one that states that workers can nap at their break, without the danger of losing their job and/or their reputation. . . . Such a policy would state that employees napping at their break is an acceptable activity, similar to other sanctioned workplace break activities such as taking a walk, eating a snack, drinking coffee, etc.[9]

Although they only briefly hint at it, the Anthonys identify the most pernicious impediment to workplace napping: the perceived detriments to one's workplace reputation. In such a situation, one's desire for sleep is held in check by assumptions about proper social behavior and the desire to appear as a productive, alert employee, even while others might be napping.

On the coattails of the workplace napping debate in the 1990s, Metronaps was launched in Manhattan in the early 2000s to offer workers an off-site place to take a nap; by 2006, Metronaps was working to establish itself as a presence in all fifty states in the United States. Metronaps was developed to service workers who lacked either the infrastructures for workplace napping or the policies that would tolerate employees napping at their desks. Moreover, as a remote place for nappers, it obviated the possible shame that they might face if they slept at their desks. As a space, it was designed to facilitate napping as well as anonymity. In the case of the former, science was brought to bear on the length of naps and the ergonomics of sleep; in the case of the latter, the space of Metronaps was carefully constructed to allow the workday napper a veneer of privacy. Located in the Empire State Building, the original Metronaps facility appeared from the outside to be a rather nondescript corporate space. Like many of the other offices in the Empire State Building, the entrance to Metronaps was a simple glass door succeeded by a front desk and a receptionist. Its lobby differed from others in having a number of stalls equipped with mirrors and hygiene products for nappers to freshen themselves with after their naps. Located behind the front desk was the napping facility, an area that hosted eight ergonomic recliners. As a space, the napping area was more akin to something out of science fiction than a cozy bedroom or a barracks: lit only with deep purple light and drenched in the sound of a white-noise generator, the black recliners faded into the darkness of the room. Speaking to the Metronaps employee who acted as my guide was difficult because of the white-noise generator, which was so overbearing (and yet not loud) that we had to huddle together to have a conversation in the napping room. She explained that the white noise was so powerful to obscure the sounds of other sleepers and people coming and going from the room; no one, to her knowledge, was unable to sleep because of it. This, however, might have had more to do with the clientele, who might have been chronically sleep-deprived and able to sleep through anything. While busy, the company's location in the Empire State Building paled in comparison to locations throughout the United States began to close in the late evening, usually around 10 or 11 P.M., only to reopen in the early morning. This prevented Kinko's from having to pay third-shift workers, who like most third shifters, earn higher wages for working an undesirable shift. To compensate for these closures, regional printing facilities have been created, where production work from a number of Kinko's locations is sent for overnight processing. In areas like metropolitan Detroit, which houses a number of corporate headquarters and automotive research facilities and where forty to fifty overnight employees were employed by Kinko's in the 1990s, only a handful of overnighters have remained, and they have been consolidated at busy or profitable stores and a local production facility. Kinko's is not an exceptional case; others include a number of overnight diner chains, fast food restaurants, gas stations, and local and national grocery stores. As might be expected, being open twenty-four hours a day is a precarious business decision and depends on balancing consumer goodwill and profitability. As Kinko's and others have discovered, the meager benefits of consumer goodwill might not be worth the losses in profits, and maintaining normal business hours preserves ideas about respectability founded in American normative ideals, in addition to obviating the need to pay third-shift workers.

Employees in India-based call centers experience even stranger things with their work schedules than Americans do. Rather than being normalized along a 9 to 5 schedule in their own time zones, they are scheduled according to business activity in the United States. A call to India at 9 A.M. in the United States' eastern time zone coincides with early evening in most parts of India; as a result, most Indian call center workers work the third shift, beginning in the late evening and working through morning. While most employers hire employees to work eight-hour shifts, most work demands (e.g., the number of customers consulted) force employees to stay at their jobs for twelve to thirteen hours. As a result, many employees experience the traditional by-products of shift work, including insomnia and chronic sleep deprivation. Moreover, Indian work schedules are synchronized with U.S. holidays, and employees there receive July 4th, Presidents' Day, Christmas, Easter, Labor Day, and New Year's Day off, but they are required to work on Indian holidays. This sort of spatiotemporal imperialism is a subtle one, and its effects have yet to express themselves fully. The lives of these Indian workers become desynchronized from those of their families and friends outside their workplace, for the benefit of being coordinated with Americans who seek help. Stranger is the jetlag benefit that U.S.-based managers receive when traveling to international call centers: they can continue to sleep as usual and conduct their business at local nighttime hours. Thus, rather than providing a pharmaceutical cure for jetlag, American businesses may be arranging temporal fixes and a globally available spatiotemporal order, predicated on the powers of normative ideals of sleep, for some.

The coordination of work times—locally and globally, individually and institutionally—invariably requires the coordination of society and biology through regimes that define and legitimate normative models of everyday life. The demands of American capitalist formations require some workers to work at undesirable hours, personally managing their desires to achieve balance in their everyday lives. But if genetic fantasies such as that of Walsh come to fruition, more nuanced models of everyday life may be institutionalized, which, instead of relying on such graceless coordinations as shift and night work for witless laborers, would depend on more intimate biological knowledge of workers and find means through economic rationales and social obligation to coordinate spatiotemporally appropriate bodies and labor. The intensification of public interests in the desire for sleep should be seen as partly motivated by capitalist interests in producing working bodies, bodies resistant to workplace fatigue and tirelessly alert, an objective developed since the beginning of the industrial age. Although Walsh's genetic fantasy is unlikely to come to full fruition, the fantasy of a biologically discerned workforce directed by government agencies may produce the expectation among employers for workers to intimately know their desires and select appropriate work times accordingly. Alternatively, workplaces may become more explicit in their arrangement, charging workers with regulating their sleepiness and alertness through institutionally supported means. This may lead to the incorporation of sleep-related work benefits that will be used to attract and retain workers; if given a choice, workers who know themselves well as sleepers will always find a sleep-friendly workplace preferable to a sleep-antagonistic one.

10

····················

Take Back Your Time
Activism and Overworked Americans

A BROAD ATTEMPT TO CONFRONT THE CAUSES OF AMERICAN
sleep deprivation has come in the form of the Take Back Your Time
(TBYT) movement, a loose network of academics and activists who have
drawn attention to the persistent overworking of Americans. They take
as their object of criticism the expansion of the American workday; to
draw attention to the systematic overworking of Americans, they have
attempted to establish a national holiday: according to their designs,
October 24 should be designated as Take Back Your Time Day. The
symbolism of October 24 is important, as one might expect: measured
against every other industrialized nation in the West, the amount of time
that Americans are overworked is represented by the span of time between October 24 and December 31. In the introduction to the TBYT
manifesto, sociologist Juliet Schor argues:

> The average worker in 2000 could produce nearly twice as much as
> in 1969. Had we used that productivity dividend to reduce hours of
> work, the average American could be working only a little more than
> twenty hours a week. . . . Taking all productivity growth as leisure
> time would have led to a stable real level of income.[1]

With this foundation, other contributors to the TBYT manifesto argue
for increased family time, greater vacation time and sick leave from work,
maternal and paternal leave for new parents, and an overall reduction in

work time. What has become a central point for the elaboration of the TBYT movement's critiques is the role of the family in American society. This, unfortunately, naturalizes some romantic ideas about the ways families might and should relate to one another as well as the normative spatiotemporality of family relations. But, in the face of American impulses toward the encroachment of the workplace into family time, this focus is a deliberate choice and may appeal to both beleaguered parents and their (possibly neglected and overextended) children.

Each year the TBYT movement identifies a site of poverty in contemporary American use of time as a locus for activism. In 2006, the campaign slogan was "Get back to the table!" and emphasized the benefits of family or communal meals and activities. In that year's annual press release for Take Back Your Time Day, this new emphasis was explained:

> The family dinner table stands empty. The card table collects dust in the closet. Cafes serve their drinks to go. In response, the US/ Canadian Take Back Your Time campaign is calling for Americans and Canadians to "get back to the table!" this October 24th. . . . "Time for family meals has disappeared as parents work long hours and kids scurry to activities intended to beef up their college applications," says William Doherty, a professor of family studies at the University of Minnesota. Dr. Doherty, spokesperson for the *Reclaim Dinnertime* campaign, also points out that "shared family meals are a leading predictor of how well students do in school."[2]

This "Reclaim Dinnertime" campaign was cosponsored by Panera Bread, a national bakery in the United States. While this was a noble pursuit, and it was shored up with a host of scientific studies that evidence the beneficial effects of family or group meals, Panera proposed that families "on the go" might use its restaurants as a dinner room away from home: "Panera recognizes that many 'on the go' families today do not have the luxury of gathering around the dinner table at home every night of the week. So, if you are looking for a place to reconnect outside of the home, we invite you to make Panera Bread your gathering place."[3] Like the recasting of the siesta as the workplace nap, the family dinner is being capitalized, transformed into a consumer good. As the Reclaim Dinnertime Web site demonstrates, the movement could succeed only if it could also become a fashionable object and an appealing means to settle individual and family experiments with social uses of time. At best it could found a new form of life alongside reconceived intimacies and desires.

At the 2005 biannual meeting of the TBYT movement, a flier was distributed that suggested a number of "time tweaks": "large and small changes, adaptations, and innovations in policy, business, society, and daily life that can help workers, parents, students, and families reclaim their time!" Two tweaks presented were grand in their scope and debatable in their effects. The first was "softening the standard time–to–daylight time transition." This echoed the original plans of William Willett, the father of daylight-saving time, who suggested making the transition between standard days and daylight-saving days a gradual one. Willett's plan was derided as confusing, however sensitive to the desire for sleep it might be. The TBYT pamphlet suggested:

> Half-an-hour-later start times of events, meetings, classes, and work shifts in the six to eight weeks that follow the . . . transition from winter's Standard Time to summer's phase-advanced Daylight Time schedules . . . and during the six to eight weeks that precede late autumn's return from Daylight back to Standard Time . . . would make life easier.[4]

It was fully expected that not all individuals or institutions would endorse such tinkering, but it was suggested that "the weeks following [the transitions] should be a period of maximum possible Flextime." The second tweak was a proposal for a thirty-six-hour workweek, which would consist of four nine-hour workdays. The immediate benefit identified was "one-fifth less commuting," which would "reduce traffic congestion, carbon dioxide emissions . . . and public and personal costs of all kinds." The officials of the TBYT movement recognized that "business, the public sector, labor unions, other institutions, and policymakers" would need to agree on such a radical shift in the revision of the spatiotemporal order of everyday life. Such a utopian plan would seem to fly in the face of the incessant rhythms of contemporary society, which would require substantial restructuring in order to accommodate a broad redevelopment of the workweek, however appealing such a reformation might be, both individually and institutionally. Effectively, a new hegemonic spatiotemporal formation would have to be developed for American society—no small feat and potentially as disorder producing as the contemporary formation.

The Take Back Your Time movement found ready allies in some faith-based community groups. This might be rather ironic, if one accepts Max Weber's hypothesis that it was Protestant ideas about work and productivity that established American capitalism as in ever more need of workers' time. In 2006, the Massachusetts Council of Churches, a statewide ecumenical organization, endorsed a program to encourage their congregants to take "four windows of time" between Labor Day and Take Back Your Time Day "to rest and recreate balance."[5] To shore up the council's assertions of the necessity of reclaiming time from work obligations, a quote from the New Testament was included on a promotional poster for the project: "The apostles gathered around Jesus, and told him all that they had done and taught. He said to them, 'Come away to a deserted place all by yourselves and rest a while.' For many were coming and going, and they had no leisure even to eat" (Mark 6:30–31). This appeal to the orthodox logic of the Bible is echoed by a blurb created for the poster, tucked in its upper left corner: "God took seven days to create a perfect balance, including time for rest. Our society is out of balance." In a flier describing the intent of the project, the authors suggested, "Commit at least four windows of time—ideally a consistent time each week for four weeks—for simple, restorative activities. Take time to be with self, God, family, friends, community, nature—time to restore your soul." The flier goes on to rule out "scheduled activities," "buying or selling," "stress," "intrusive technology," "obligations," "work," and "guilt." Curiously, this push from religious officials stemmed from a recognition on their part that "clergy are among the most overworked professions in America." This was included as part of a flier on "time as a social justice issue" and acted as a slightly shaming device, asking readers, are you "sensitive to the need for rest and renewal for clergy?" The rigors of the Protestant ethic, it seems, capture even those who are formally divorced from the strict demands of the dominant American capitalist spatiotemporal orders.

Although the American workday might incorporate flexible work schedules and workplace naps, by the turn of the twenty-first century neither of these options had managed to grossly restructure dominant spatiotemporal formations of American institutions. Instead, twenty-first-century American spatiotemporal formations align with historical and cultural models of productivity and respectability: the 9 to 5 workday, now supplemented with pharmaceuticals. Moreover, on the global scale, coordinations between distant places have relied not solely on communication technologies to bring people together but also on spatio-temporal coordinations deployed by those in power to preserve benefits at the expense of those workers in remote places, an effort facilitated by global disparities in work and pay. Although some institutions have found ways to meet their workers' desire for sleep, these experiments are solitary ones, despite the formal acceptance of flextime and workplace napping by some employers. A workers' revolution on the order that the TBYT movement calls for is unlikely, but its promise lurks in the corners of contemporary American politics, offering the possibility of a new spatiotemporal formation of the everyday and a future unlike that imagined by any of the workers' movements of the nineteenth and twentieth centuries. American spatiotemporal formations in the twenty-first century draw increasingly upon one's ability to know oneself as a sleeper and an understanding of the desires of others.

Child patients who were brought to the MSDC invariably confronted both physiological problems and social disorders, the latter generally evidenced in an inability to attend school at the expected times or an inability to stay awake through the day. These situations often compounded the difficulties faced by young patients, resulting in social estrangement, academic underachievement, and institutional struggles. Largely because of this spreading disorder in the lives of young patients, the clinical staff at the MSDC were often compelled to find means to insinuate their authority into educational institutions to ease the situations of students. In one case, Dr. Richards introduced Rick, a teenage patient who was rather exceptional in that he actually required eleven hours of sleep per night, one of the highest amounts ever recorded at the MSDC for a teenager. Rick was having problems at school; Richards remarked that he just "can't go to school at 7:30 A.M.," which led to the assembled staff's short discussion about the intolerability of the educational system. They decided that, at its base, the rationality of institutions was such that "either you're there at 7:15 or you're not there at all." The resolution of the discussion focused on what the sleep center physicians could do in order to ease the patient's problems at school, looking at clearing his absences and tardiness with their authority, but also trying to arrange some sort of home-schooling situation with the parents.

Should it be medical practitioners who lead a charge on the rethinking of school start times and the flexibility of school in American society? the flexibility of work and work times? Although the physicians at MSDC might have staged limited interventions on a case-by-case basis, a more expansive critique of the start times of American schools found itself foundering against cultural expectations of normative uses of time and space: the agrarian myth. One example of successful change has been the attempt of the Minneapolis-based Center for Applied Research and Educational Improvement (CAREI) to shift school start times. This shift was accepted on a local basis and was weighed against the intimate connections of the community, judging the impacts of school start times on students' bodies against the inevitability of modern spatiotemporal and social obligations of parents whose lives depended upon the rhythm of school times.

Throughout the United States at the end of the twentieth century, middle and high schools tended to begin between 7 A.M. and 8 A.M., with elementary schools generally beginning between 8 A.M. and 9 A.M.[6] As children age into adolescence, their desire for sleep changes, from an average of about eight hours per night to upward of nine and a half.[7] As most American parents can attest, the number of distractions in a child's life parallel this increase in sleep demands; by choice or necessity, the social obligations of a teenager are generally more time consuming than those of elementary-aged students. This increase in distractions decreases the chance that teens will go to sleep at a time that will accord them the nine and a half hours of sleep they desire for optimum attention throughout the school day. The University of Minnesota sponsored a study in local school districts, which was spearheaded by CAREI, and it asked high schools to begin anywhere from an hour to an hour and a half later than usual. The results produced by this study were positive: student and faculty attitudes cheered, grades improved, standardized test scores increased, and the incidence of the symptoms of attention deficit disorder (ADD) and attention deficit hyperactivity disorder (ADHD) decreased. Although direct causality between increased sleep and these improvements is not claimed by CAREI, recommendations were made in the form of a report to the U.S. Department of Education, as well as by local school boards that called for later school days—and CAREI mustered the statistical evidence to prove it. Incidentally, this same study also showed that previous to the change in school start times, most teenagers averaged about seven and a half hours of sleep per night.

The school day as it was developed and concretized is based upon parental work times, because early public education was a means to provide child support for the growing working and middle classes.[8] Throughout the nineteenth century, public education of this form—as a means to provide parents with time in which to work, free of their children—was a necessary economic function provided by the state. The structure of schooltime was thus necessarily related to work times, with schools being open for parents to leave their children for the duration of their workday. With the turn of the twentieth century, the development of the shorter school day occurred concomitantly with the shortening of the workday, the spread of the automobile, and the stabilization of the American middle class. With the further suburbanization of American society, the timely transportation of students became a primary concern, with public busing services ensuring that students were delivered to school and returned home without interfering with commercial and commuter traffic. This simultaneously ensured that a period after school was provided for students to pursue extracurricular activities, including school sports. As detailed in the CAREI reports, the primary stumbling block to the transformation of school start times was school sports at the high school level, an issue that weighted the needs of the few school athletes over the desires of the rest of the school's population. Only when the timing of sporting events could be altered throughout an entire district could the school day be shifted in such a way as to provide enough sleep time for the entire student population. This kind of shift usually requires elementary schools to trade their later start time for the earlier start time of the high schools, and although the shift preserves school start times as a whole, it exchanges the ages among those start times as deemed appropriate. As such, this kind of shift may still abide by the agrarian model and industrial demands of the early school day, but other possibilities for arranging the school day are available, if not readily pursued.

One possibility is that of a flexible school day, not unlike the flexible workday imagined throughout the 1990s. This arrangement was suggested by the physicians at the MSDC, who frequently saw school-aged patients who faced difficulties in not being able to attend or fully participate in school. Dr. Pym, one of the senior pediatricians on staff at the MSDC, considered the gambit of problems associated with school start times and patient disorders in his discussion of Dennis. At the time of Dennis's clinical assessment, he was sixteen years old. As a result of his school's schedule, he had constant problems with daytime sleepiness, which led to troubles he faced with truancy agencies as well as the school. Luckily the school was willing to be adaptive, allowing the boy to come into school when he could, at around 10:30 A.M. His performance improved remarkably as a result, but his parents and teachers still had some concerns, because he fell short of credits to graduate on time, requiring that he stay in school longer in order to earn his diploma. One of the other physicians inquired as to the school: What school would be willing to let a student come in nearly two hours late? Pym explained that it was a high school in one of the surrounding small towns, possibly more amenable to such accommodations because of its size. In contrast to Dennis's school, which allowed a flexible spatiotemporal order, the institutions and actors that shape the American day tacitly accept the flexibility of human desire for sleep. With the inevitability of the school day, as well as the school week and year, students are left to develop strategies to balance desire and obligation.

In the following excerpt from an interview with Kathleen, a white female who had recently been diagnosed with idiopathic hypersomnia, restless legs syndrome, and a nebulous circadian rhythm disorder, the complexity of desire and obligation that many students face is highlighted in her consideration of what is normal for teenagers and how individuals, not institutions, might deal with it when faced with inflexible institutions:

> I have always been someone who is unusually tired all the time, but when I was in high school I kind of thought that was normal, because teenagers are supposed to be tired (I guess is what I thought). And before that I just never really knew it [my sleepiness] was abnormal. I was a teenager, and my schoolwork was predominantly reading and writing, and fortunately I would be able to go to the library for a chunk of hours and take a nap for twenty minutes or so. . . . I've always had caffeine; I don't like coffee, but I drink soda

at least every day. So I didn't think that affected my sleepiness, so even if I tried staying awake during high school by drinking coffee, it never really worked.

This recourse to stimulants is necessitated by institutional inertia that values the tradition of schooltimes and parental expectations over the desires of students. Invariably, then, as students negotiate their intimacies and desires, they are led to ingratiate themselves into the therapeutic milieu of American society, wedding social obligations and cultural expectations with economic demands and chemical investments that offer temporary remedies without curing the fundamental social structures that promote their poor sleep.

The institutions that compose our everyday lives lay the basis for our desires: for work and play, education and care, sleep and wakefulness. When our desires for social interaction are disrupted, whether by our physiological demands or by the inflexibility of society, disorder arises. As long as everything proceeds smoothly, in an orderly way, there is no call for medical intervention. But this existing state of affairs can obscure what individuals have to do in many cases, which is muddle through their lives by supplementing their sleep with caffeine and more illicit substances. When we make the choice, conscious or not, to fight off a feeling of sleepiness with a cup of coffee or another stimulant rather than taking a brief nap, we also tacitly support the dominant order of sleep and wakefulness in society. There are other everyday orders that Americans might invest themselves in, more flexible structures that make demands of institutions to accommodate individual variation in the need for sleep rather than require individuals to meet strict spatiotemporal demands. And, as the TBYT movement makes clear, there are broader reorganizations of society that we might consider: Must we work as much as we do? And sleep as little or as erratically as we do? One of the powers of integral medicine is its ability to return the disorderly sleeper to orderly society; however, with a more flexible social form, this integral function may be less necessary. We may come to identify ourselves less through medicine and more through our understanding of ourselves as sleepers. This may simply substitute one biopolitical order for another, replacing medicine with science, or it may initiate some other way to conceptualize our desires and intimacies that is yet to be discovered. Hints of the possible futures of sleep lie in the various limits of sleep that have been considered over the last century, with roots much further back. Conceptualizing the futures of sleep necessitates turning toward fiction and fact, science and history, especially as they are meted out through the law, science, and medicine early in the twenty-first century.

III.

THE LIMITS OF SLEEP

11

············

Unconscious Criminality
Sleepwalking Murders, Drowsy Driving, and the Vigilance of the Law

A London court yesterday sentenced a man to jail for having "willfully disturbed" other persons by snoring in an air-raid shelter. . . . What is interesting is the court's decision that the snoring in question was "willful." In reaching this conclusion the magistrate overruled the offender's seemingly plausible plea that "I can't help what I do in my sleep." Does this ruling imply that all snoring is "willful," or did some special, unreported circumstance apply in this case? If there is a distinction, how does one tell the difference between a willful snorer and a nonwillful one?

—*St. Louis Star-Times* and *Chicago Daily News*,
December 18, 1941

ARE WE RESPONSIBLE FOR THE THINGS WE DO WHILE WE SLEEP? If we know that we are restless sleepers, snorers, sleepwalkers, or otherwise disorderly, are we liable for the actions we commit while unconscious? If there is a treatment for our disorderly sleep and we refuse it, are we liable for that decision? These questions of culpability are at the center of American conceptions of agency, the motive power of individuals, and desire: What motivates our actions? Sleep is especially troubling because even we cannot know ourselves as sleepers; we know our subjective experiences of sleep, but these can vary dramatically from what others tell us about our sleep. How we sleep is often beyond our control, although we might be able to control when and where we sleep. Snoring, restless limbs—those we might be able to mollify through therapeutics or social arrangements. But what about sleepwalking and other things we do while asleep? The history of sleepwalking murders opens up these questions of culpability and agency, as does the more recent history of "drowsy-driving" laws. Americans have long been struggling to conceptualize the extent to which individuals are culpable for the actions they commit while unconscious, and by examining the various moments of conflict around these issues we can see how ideas about agency and sleep have changed over the last century, ideas whose roots extend back to the eighteenth century. These questions of culpability open up futures of sleep: If chemical consciousness is accepted as normal, what are the repercussions for those who act unlawfully while under the influence of medicine? for those who are noncompliant with medical therapies? for those who use sleep as a defense for illegal actions?

Take, for example, the cases of drowsy-driving accidents and sleepwalking murders. If individuals fall asleep while driving and cause injury to others, they are more likely to face prison time than if they commit murder while sleepwalking. Sleep is particularly difficult for the American legal system because it muddles ideas about behavior and biology in attributing culpability and desire to individuals: falling asleep is perceived as willful and thus calculable within the terms of the law, whereas behavior during sleep is accepted as unruly and imbued with happenstance and the unpredictability of nature. In the following I examine how sleep is understood as biological (specifically as it relates to drowsy driving) and as behavioral (particularly in the influences waking life has on events during sleep), and I look at the various ways the human nature of sleep is deployed within the law to formulate models of desire and culpability. If sleep's sovereignty is accepted, as William Dement argues for, then the status of sleep's effects as sovereign must be accepted. What one does while sleeping, in other words, is beyond the control of the sleeper. Sleep, in this figuration, is akin to bodily possession, much in the same way that dreams are sometimes construed as being mediums of external forces. Sleep unsettles the law, providing a means through which integral medicine can insinuate its powers into conceptions of biology and society, making individual desire into the desire of the masses.

Vehicles of Destruction: Edgar Huntly at Large

In 1799, with his depiction of tormented Edgar Huntly, Charles Brockden Brown offered a preliminary figuration of the sleepwalker who commits murder, a figure in many ways that haunts contemporary American ideas about law and the validity of the "sleepwalking defense." *Edgar Huntly* is often read as a colonial novel concerned with the relationship between the new colonists in the Americas and what was perceived as untamed nature, both in the Native American populations and in the landscapes that made up the American environment—a nature, it might be argued, that lies dormant in every "civilized" human. Sleeping and dreaming are taken by Brown and his protagonist to exist on the same relative plane; one is intimately tied to the other, and both represent an inevitable, natural force that compels Edgar to commit extreme and unlawful acts under the influence of sleep. But the sleepwalker is understood by Brown to be porous, as is exemplified by Edgar and his double, Clithero: they attend to the world around them, sometimes in violent and unpredictable fashion and always amnestically. When Edgar sleeps, he is influenced by his surroundings as well as the narratives that he is told. To be fair to poor Edgar, whom so many tragedies befall because of his sleeping problems, he is a sensitive young man, prone to external influences: it is not his intent to commit the acts that he does, but rather he becomes caught up in the force of dreams, sleep, and their disorders, the results of which are devastating to him and his world.

For the American legal system early in the twenty-first century, sleeping bodies are problematic because of their lack of consciousness and willfulness and their ability to inflict harm on themselves and others. This dilemma of culpability stretches back to the misadventures of Edgar Huntly. The questions of culpability play an important role in the use of the sleepwalking defense for murderers and the implementation of drowsy-driving laws. The order of everyday life offers a patina of anonymity; when the order is broken, when it becomes disordered, culpability is assigned to those who unsettle it. The masses offer anonymity and innocence, but when individuals are identified as the source of disorder, culpability is attributed to them to make sense of their disorderly acts. But willfulness complicates attribution, especially in the case of crimes related to sleep, because the intent of an individual is difficult to discern, entrenched as it is in the formation of everyday life. Edgar Huntly is the preeminent vehicle for sleep's destructive forces, unsettling the law through his sleepy unruliness. Being unruly in his way, he is an archetypal example of the dangers of desire, for accepting Edgar as willful is to accept the terrible possibilities that he embodies as within the realm of the normal.

Returning home from time abroad, Edgar passes the gravesite of a close friend who was murdered years earlier under mysterious circumstances and whose killer has never been found. Eerily, at the gravesite stands a man, digging. Edgar watches from afar, eventually discerning the identity of the man and then supposing that he has deduced what befell his friend years ago; he imagines that the guilty murderer has returned to the grave of his victim. Unwilling to confront the suspected killer directly, Edgar circuitously comes to learn from the man's acquaintances that he experiences what would now be recognized as dream-enactment behaviors: in the middle of the night, he often leaves his bed to reenact events that happened years earlier, but, unfortunately for Edgar, not the events that brought about the destruction of his friend. Eventually Clithero, the dream enactor, relates to Edgar his sorrowful tale, about how, because of a series of circumstances that could be found only in a gothic novel, he found himself poised over the sleeping body of his dearly beloved fiancée, seconds from killing her. What led Clithero to this? Sleep deprivation. Caught in a bind of circumstances too convoluted and too irrelevant to reprise here in their full detail, Clithero was driven to kill his matron, a woman who had adopted him as a servant and confidant years previously. In a fugue brought on by guilt-induced sleeplessness, Clithero decides to kill her to save her from the traumatic knowledge that her treacherous brother was alive for years while thought dead, only to have been recently killed accidentally by Clithero's own hand. Clithero then decides to creep into his matron's bedroom to stab her, but his own indecision stops him at the last moment. His matron has given her bed for the night to her daughter, Clithero's betrothed, and she looks on as Clithero is about to murder her daughter. At the final moment, she interrupts his fugue, and Clithero flees, leaving Great Britain for the American colonies. Hired by a local aristocrat, Clithero lives his life with the guilt of his previous actions and is unable to face them or the people whom he has wronged; by night, he flees into the woods while asleep and enacts his dreams.[1] In so doing,

Clithero is the epitome of the model of sleepwalking whose emergence is attributed to the power of suppressed thoughts, a recurrent motif in American narratives of sleepwalking.[2] This is what Edgar stumbles across, and it eventually leads to his own somnambulism through some sympathetic contamination: the idea of this kind of sleep leads Edgar to sleep the same way, albeit unwillingly.

Shortly after his discussion with Clithero, Edgar awakes one day to find himself transported from the bed he fell asleep in to a cave in the wilderness.[3] Edgar has no recall of how he came to be in the cave, and after shaking off a bout of what appears to be sleep paralysis (the inability to move coupled with a vague sense of dread), he sets about trying to find his way from the cave. What follows is a series of misadventures that result in his bringing about the destruction of a handful of colonists and natives and being chased by townspeople brandishing farm tools as weapons. To say that Edgar is an unreliable narrator is to grossly misunderstand him: the results of his sleepwalking episode read more like a fevered dream than accounts of actual events, with Edgar handily dispatching opponents with single shots of a musket and fatally wounding them in knife fights—quite a feat for a young, effete aristocrat. His strange fantasy adventure reads like the narratives of later sleepwalking murderers, and it establishes, indirectly or not, the genre of sleepwalking storytelling that haunts American law. Any inconsistency in Edgar's narrative is to be taken as evidence that he was indeed sleepwalking, that he was only marginally aware of the waking world while somnambulating. Maybe Edgar did not kill anyone; maybe it was all a dream. Events appear as disconnected, with Edgar waking to find himself in beds he did not lie down in, circumstances that are then taken by him as further evidence of his sleepwalking. By the end of his narrative, it is impossible to discern what might have actually occurred and what has been sheer fantasy on Edgar's part. This results in the destabilization of the narrative (did he ever meet Clithero, and was the latter's tale a fantasy as well?) and in the eventual, anticlimactic collapse of the novel. As a gothic novel and a story about sleep, *Edgar Huntly* is a perfect example of a recursive narrative, a narrative that destabilizes itself as it attempts simultaneously to assert its validity. These tropes are engaged throughout the nineteenth and twentieth centuries in the use of the sleepwalking defense, but they are catalyzed in this novel quite succinctly.

As mentioned above, the other problem with Edgar is that he is a porous, or sympathetic, sleeper, and he mimics or reenacts what other people tell him about their own sleep. This happens in the case of Clithero's confession to Edgar, the latter seemingly the former's one and only confidant. The novel reads as if Clithero infects Edgar with somnambulism, because it is only after their contact and Clithero's confession that Edgar has a bout of sleepwalking. If Edgar is an unwilling sleepwalker, assuming for the moment that his narrative is true, both in the sense that sleepwalking is some alien affliction given to him by Clithero and in the sense that his specific actions while somnambulating are unintentional, then he could be doubly vindicated by the law. But the villagers who pursue him see things quite differently, hence their pursuit of him with weapons. Edgar is fortunately able to extricate himself from both the proclivity for sleepwalking and the vigilante justice of his fellow colonists, but, in so doing, he posits the foundation for the sleepwalker as victim or the sleepwalker as sovereign in the face of the law, a presumption that would recur throughout the nineteenth and twentieth centuries. To be isolated from the masses, to be individuated, is necessarily to be identified as culpable, to be willful; aligning the actions of the individual with expectations of nature is to return the individual to the masses. This is especially evident in the case of sleepwalking murders, where the alignment of individuals with nature absolves them of their willfulness and mollifies the effects of their actions by rendering them part of nature, simply biological. Nature, however, can be both a justification and an excuse, as the iterations of the sleepwalking defense make evident.

A History of Unconscious Intent: Sleepwalking Murders

The history of the sleepwalking defense in the United States is a perplexing one, having as much to do with the personalities involved as with the changing conceptualizations of sleepwalking. The first well-known case of the sleepwalking defense was that of Albert Tirrell in the murder of Maria Bickford, in Boston in 1845. Tirrell was tried not only for the murder of Bickford but also for the attempted arson of the room where her dead body lay (a room that they shared briefly in a lodging house) as well as for adultery, because he was already married during his confirmed affair with Bickford. He was acquitted of the murder and arson charges on account of his sleepwalking but was imprisoned for three years for his adulterous behavior. Bickford's body was found by the landlord who rented the room to Tirrell and Bickford,[4] after he had been awoken by the victim's screaming early one morning and the sound of a person running down the home's stairs and fleeing the premises. Bickford's throat had been cut "nearly from ear to ear." The police found a bloodstained razor and a washbasin containing bloodied water in the room with Bickford's charred body; they also found some men's clothing that had been left behind, as well as a letter to Bickford with Tirrell's initials on it. The question was not whether the murder had been committed by Tirrell but whether he was conscious and therefore responsible for his actions. At the time of the murder, Tirrell had already been arrested for adultery with Bickford and had been released on bail under the supervision of his wife and family, who hoped to reform his lascivious behaviors. It was during this period of parole that he committed the murder and fled Massachusetts. Tirrell traded on the sympathies of his jurors and was able to provide them with a plausible means for absolving him of guilt, namely, somnambulism.

Tirrell was acquitted of the murder and arson; in contemporary American society, he would most likely have been found guilty of both the arson and the murder, as these actions were quite clearly willful according to the evidence at hand. Upon exiting the premises of the murder, Tirrell found his way to "a livery-stable near Bowdoin Square," saying that "he had got into trouble; that somebody had come into his room and tried to murder him."[5] He then sought refuge with his wife's family, who (under what expectations of his behavior are unknown) gave Tirrell funds enough to flee to England. The ship he boarded was turned back because of inclement weather. Tirrell then boarded a ship to New Orleans, where he was intercepted by local law enforcement, who had been tipped off to his eventual arrival. Tirrell was extradited to Boston, where he was received by a public fervently awaiting his trial: rumors of Tirrell's actions gave grist to the Boston press, and stories about the murder and Tirrell's flight consumed the public imagination. If Tirrell's trial was a celebrity trial of the nineteenth century, then his lawyer, Rufus Choate, was partly to blame: by all accounts, his boisterous and charismatic presentation, as well as his ability to induce doubt about even the most explicit evidence, swayed the jury to acquit Tirrell. Or, quite possibly, the all-male jury of the day was more willing to detest the actions of Bickford as an immoral woman than they were willing to find the son of a well-respected local merchant guilty of the woman's murder. The sleepwalking defense might have simply given the jury a justifiable way to find Terrill innocent of both the murder and the arson.

In the diary of a contemporary, John Langdon Sibley, one finds a clear explanation of the circumstances of the verdict; the entry dated March 28, 1846, reads:

> The excitement in Boston caused by the trial of Albert J. Tirrell for the murder of Mrs. Bickford has been brought to a close by the verdict of Not Guilty. The apparently novel ground of Somnambulism was introduced and strongly urged in his defence; but the jury acquitted him, without even mentioning Somnambulism in their consultation. The tone of public sentiment is such in regard to capital punishment that it is very difficult to convict a person for a capital offence; & when such a conviction takes place, public sentiment demands a commutation to imprisonment for life. General opinion is that Tirrell is guilty; but it would have been unreasonable to have convicted him, upon the evidence adduced.[6]

This successful use of the sleepwalking defense spawned its subsequent use throughout the eastern seaboard states, but it was quickly regarded as an untenable explanation for the variety of crimes it was employed to explain. However, one notable case of another acquittal based upon sleepwalking was that of Simon Fraser, who, disoriented upon awakening from a nightmare, dashed his young child against the floor of the room he and his wife shared with the toddler. He was acquitted for his clearly unwillful actions.[7] In this case, Fraser clearly had no reason to murder his child, and sleepwalking made sense of his aberrant behavior, whether or not he had the sympathies of the jury. Because his desires could be made plain, because his intimate connections were so apparent, his culpability was assuaged. Later, after a century of quiescence, the sleepwalking defense returned with full force in Arizona in 1981, when Steven Steinberg was accused of murdering his wife.

The trial of Steven Steinberg focused on the behavior of his wife, Elena, rather than on his own. She was portrayed as insensitive and domineering,[8] and in this way, the trial of Steinberg echoed the trial of Tirrell, placing the victim on trial rather than her killer. In Steinberg's defense, his lawyers, drawing on lay psychoanalytic understandings of the unconscious, argued that his suppressed rage at his wife eventually leaked out during his sleep one night, bringing him to murder her in her sleep. Central to understanding Steinberg's culpability were his social desires and troubled intimacy with Elena, which revolved around his interests in gambling. Steinberg had stabbed his wife twenty-six times with a kitchen knife, claiming no memory of the attack. He first claimed that two bearded men had broken into the home, had attacked him and his wife, then attempted to rob them, and were eventually chased off by Steinberg, but not before fatally wounding Elena. The investigating officers immediately suspected this as a lie: Steinberg had a single wound on the palm of one hand, while his wife was riddled with stab wounds. Moreover, for violent thieves to attack a small woman in bed rather than her husband seemed an odd choice in the commission of a burglary. The investigation revealed that Steinberg had a long history of attacks by similarly bearded, wild men, who provided ready explanations for the money he lost while gambling. He also had a long history of a gambling problem, as well as embezzlement, and just before the murder was in need of yet another bank loan to pay off his accrued gambling debts. On his wife's death, her life insurance would pay out with him as the beneficiary.

To found Steinberg's sleepwalking defense, his attorney brought in two psychiatrists, who claimed that Steinberg was a sleepwalker. (Recall that this was the early 1980s, and little in-depth research had been conducted on sleepwalking.) This claim was based largely on the testimony that his two daughters also somnambulated rather than on any evidence of his own sleepwalking. His attorney also argued that sleepwalkers could exhibit complex behaviors. In this case, Steinberg had to walk to his kitchen and back to the bedroom, stab his wife repeatedly with a nine-inch blade, conceal the blade beneath the bed's mattress, throw some clothes from the dresser drawers around the room in haphazard fashion to make it appear as if burglars had ransacked the room, and then call the police. Also while sleeping, as attested by his adolescent daughter, he said to that daughter, "Shut your fucking door," when she asked what her mother's screaming was about.[9] All of the behaviors are, indeed, complex for a sleepwalker and could have been taken as evidence of his consciousness. But the jury disagreed with the latter interpretation and acquitted

Steinberg, who then left Phoenix and the lives of his two daughters. Steinberg's acquittal might also have had something to do with the directives of the presiding judge, who offered first-degree murder or full acquittal as the only options for the jury's finding; they might have found Steinberg guilty of a lesser degree of homicide given that choice. But, as in the case of Tirrell, the sleepwalking defense was deployed not so much as an excuse for the actions of Steinberg as a convenient way for the jury to find him not guilty; the jury may have been more inclined to sympathize with Steinberg than with his "Jewish American princess" of a wife and her emasculating behaviors as portrayed by the defense.

The sleepwalking defense returned in 1987, this time in Toronto, Ontario. Although tried in Canada, Kenneth Parks's case became a focal point for the articulation of the modern sleepwalking defense, with sleep professionals in the United States and elsewhere working to elaborate conditions under which something as uncanny as a sleepwalking murder might reasonably take place on the basis of contemporary understandings of the science of sleep.[10] Parks, a young husband and the father of a five-month-old daughter at the time, was tried for the murder of his mother-in-law and the attack of his father-in-law in their home, twenty-three kilometers from where Parks and his wife lived, in suburban Toronto.[11] In a case of extreme but not unheard of sleepwalking behaviors, Parks drove the twenty-three kilometers between his home and that of his in-laws, ascended their stairs, and attacked his mother-in-law with a bludgeoning instrument and then a kitchen knife, stabbing her to death. He then choked his father-in-law into unconsciousness after the latter awoke and attempted to stop Parks. Parks roamed the house afterward, as attested by the teenaged children of his parents-in-law, who still lived in the home and reported hearing his "animals noises." These were later interpreted as his calling out to the young adults in an attempt to protect them.

Parks then left the home and awoke sometime during his drive back home, at which point he redirected himself to a nearby police station. He presented himself to the officer on duty at the station's front desk, bloodied and wounded; he had lacerated both of his hands during the stabbing of his mother-in-law, a point that would prove critical in his successful defense. Parks's case was a complex one and most likely the only case of a legitimate use of the sleepwalking defense, unlike those before him, in which sleepwalking provided an ostensible justification for finding the defendant innocent of otherwise heinous crimes. This led to the elaboration of sleepwalking-induced action that reinforced the validity of this defense:

> As well as the sleepwalker being incompletely aware of the environment, it is in general not readily possible to deflect the person from the ongoing behavior, which appears to be essentially preprogrammed and having to "run itself out." There is no evidence that a somnambulist during sleepwalking can either execute a conscious intent from prior wakefulness or can create an intent.[12]

Although it is impossible to ascertain what dominated Parks's conceptions of his world on the night of the attacks on his in-laws, the details of his condition previous to the attack reveal a depressed and unsettled character for whom a sleepwalking defense might have been reasonable.

Firmly working-class, Parks had finished his compulsory education and taken a job as an electrician. He had recently been fired as a result of his embezzling funds from his employers in an attempt to pay off debts related to an escalating gambling problem. He had struggled with gambling previously, but a windfall payoff had led to an escalating investment in horse racing, which quickly ended with him in debt. He had confessed his financial problems to his wife, and on the night of the murder, a day before he and his wife were to tell his parents-in-law of their financial situation, he fell asleep in front of the television while watching an episode of *Saturday Night Live*. By all accounts, Parks was under a great deal of stress, depressed, and experiencing insomnia; he also had a history of parasomnia behaviors, including sleepwalking, enuresis, and night terrors, a trio of sleep-related disorders that could be traced through both his maternal and paternal lines to a variety of relatives. Following the murder, he presented himself at the police station confused and wracked with remorse, reportedly saying, "I think I have killed some people."[13] What had occurred was quickly discovered, and while imprisoned awaiting trial, Parks was subjected to a variety of psychological tests, all of which confirmed that he was depressed, had high levels of anxiety, and experienced parasomnia behaviors. All of these factors—his remorse and confusion about the attacks, his self-inflicted wounds, his family and personal history, and the mitigating circumstances of his social life—combined to make sense of his sleepwalking defense. His desires, his intimacies, contradicted his actions, and he was found inculpable of his behavior while asleep.

By the time Scott Falater was tried for the murder of his wife in 1999, again in Arizona, the potentials of the sleepwalking defense had been mapped out in the American legal system, in large part because of the role that sleep experts had played in the reformation of ideas about sleepwalking and the unconscious in relation to the Parks case. Even though one of the experts who had testified to Parks's somnambulism, Rosalind Cartwright, testified on behalf of Falater, the evidence of Falater's guilt was apparently overwhelming for the jury. Earlier in the evening in which Falater allegedly killed his wife, Yamila, she had reportedly harangued him to fix their swimming pool's pump. According to Falater, he dutifully set about fixing the pump, albeit asleep in the middle of the night, at which point he was interrupted by his wife. At this interruption and while still sleeping, he lashed out, stabbing her forty-four times with a hunting knife. He then walked into the house, changed his clothes, placing those sullied with his wife's gore in a container with the knife used in the murder, and then placed the container in the spare tire compartment of his car. He then returned to the backyard, where he found his wife still breathing. He put gloves on and forced her body into the pool, holding her head underwater until she drowned. The last of these actions was observed by Falater's neighbor, who promptly phoned the police. It may have been the accumulated physical and eyewitness evidence that led Falater's jury to find him guilty; it may also have been a lack of the charisma or pathos that had influenced juries to acquit Steinberg and Parks. Whatever the cause, Falater was found guilty of first-degree murder, his motive for killing his wife remaining unclear but his willful intent quite apparent to the jury despite his claims of somnambulism.

The culpability of sleepwalkers is difficult to ascertain. As mentioned above, the dominant model of sleepwalking posits that it could be difficult to stop sleepwalkers from their behaviors, and the statement prepared by the researchers attending to Parks's case argued that his parents-in-law must have disrupted his otherwise benign sleepwalking behaviors, resulting in his attack on them. In this paradigm, the force of a program of actions that must "run itself out" is seen as inevitably leading to violence if interrupted, a view not dissimilar to that presented in *Edgar Huntly*. In this respect, Falater had the right idea in posing the sleepwalking defense; it was the evidence against his defense that complicated his case. However, there is actually little evidence that sleepwalkers display anger or violent behaviors upon being awakened, which complicates the use of this defense.[14] Seemingly in response to the Parks case, sleep experts posited an elaboration of the criteria by which a sleepwalker might be acquitted, arguing that in order for a court to find someone genuinely sleepwalking, that person should evidence prior sleep disorders, the duration of the behavior should be brief, and it should be "abrupt, immediate, impulsive, and senseless—without apparent motivation."[15] Moreover, the victim of a sleepwalking crime should have been accidentally present, any possibility of premeditation should be ruled out, and the sleepwalker should have evidenced confusion and some amnesia regarding the event. Finally, the event should have been precipitated by extenuating circumstances: sleep deprivation, alcohol ingestion, or the use of sleep aids. These criteria are wholly biological in their interpretation of the behavior, positing evidence of a history of sleep disturbances and a lack of possible porous factors, as in the case of Edgar Huntly; rather than providing an unconscious motivation for sleepwalking murders, this interpretation moves sleepwalking firmly into the realm of the neurological and instinctual. The satisfaction of these criteria is taken as evidence that the sleepwalker is a victim of nature's inevitability. The sleepwalking defense and the murders for which it has been employed have a preternatural appeal: they bring together a variety of public fantasies about instinct and intent, the unconscious and its repressions, and the spectacle of the murders themselves. But the same cannot be said in the case of drowsy driving.

Asleep at the Wheel

On the morning of July 20, 1997, Maggie McDonnell was struck by the vehicle of Michael Coleman, who, in a fit of sleepiness, had crossed three lanes of highway traffic unscathed before finally, fatally crashing into McDonnell's car. Upon being questioned by the police, Coleman confessed to having no sleep in the previous thirty hours and had also, in that span of time, smoked crack cocaine. Coleman was charged with vehicular homicide and reckless driving. His first jury was deadlocked, which resulted in a mistrial. Coleman was tried again for the same crimes, but this time his lawyer argued that since no specific laws forbid a person from falling asleep while driving, the jury should not be allowed to consider his drowsiness a factor in the crimes. Persuaded by this argument, the presiding judge instructed the jury that they were not to consider

Coleman's sleep deprivation a mitigating factor in the accident. The jury found him guilty only of reckless driving under New Jersey law and fined him the maximum allowable fine: two hundred dollars.[16] Incensed by this decision and the legal factors resulting in Coleman's acquittal, McDonnell's mother, after a bout of debilitating depression, set to work to draw more attention to the problem of drowsy driving. She enlisted the help of state politicians, who in 2003 successfully passed the first law of its kind, named Maggie's Law. This law posited considerations of sleep deprivation as a special clause within the vehicular homicide laws for drivers who had gone twenty-four hours or more without sleep before their involvement in an accident. Unfortunately, the law as written allowed no room for drowsy drivers who had slept, however briefly, in the preceding twenty-four hours, and it worked to treat only the symptoms of drowsy driving and not the cause, namely, that through a combination of social obligations and sleep deprivation, drivers are regularly driving while sleepy.

Many disordered sleepers report drowsy-driving incidents. These bouts of sleepiness are the result of untreated or exacerbated sleep disorders, and although the disordered sleepers I interviewed evidenced an above-average attention to their sleepiness while driving (many of them offering drowsy-driving stories with no prompting), the obligations of their work, school, and family life often coalesced to create circumstances of inevitable drowsy driving. Christine, a middle-aged woman with obstructive sleep apnea, cited her drowsy-driving incidents as the primary events that led to her seeking medical help for her sleep problems: "I think my wake up call was, um, twice in a short period of time I caught myself starting to drive off the road. . . . Without any warning, I didn't realize that I was sleepy. You know, all of a sudden I realized I was driving off the side of the road. And that scared me, of course." This is similar to the case of Martin, mentioned in chapter 4 and who also experienced panic attacks and who was made aware of his narcolepsy during a trip home after visiting a friend at a university about 160 kilometers away.

> Driving back on the freeway, it would really scare me, because I would doze off almost uncontrollably and be awoken by people honking because I was halfway in their lane. And I was looking for a rest stop, a place to get some coffee and chug as much stimulants into my body as possible. But just on the way there I fell asleep three or four times. . . . There was a time when I wouldn't drive at all. Some

of my panic attacks would occur knowing that it's happened before that I've fallen asleep behind the wheel. What'll happen is that I'll fall asleep for three or five seconds and then all of a sudden I'll wake up with a jolt. I'd have a panic attack when driving, just worrying that [falling asleep] could happen to me at any time.

In both of these cases, the sleep disorders of the individuals had yet to be diagnosed, and neither of them was aware that a sleep disorder was causing their near accidents. Stan, another disordered sleeper who was later diagnosed with severe obstructive sleep apnea, related, "I was driving along, and the next thing I knew the guy next to me [the passenger] grabbed the wheel, and I had no idea where he came from. And I was awake the rest of the trip." A couple years later and while working a very disruptive shift work job that constantly changed his working hours, often with little forewarning, Stan had another drowsy-driving incident: "I was coming home one afternoon, and I'm pulling into a traffic light, and I realize I'm not stopping, and I basically had to go into the left lane and drive over a traffic hump to avoid oncoming traffic. And I did it—it scared the shit out of me." Stan had learned that either he had to have someone else drive when he was feeling sleepy, or he had to stop to rest and briefly exercise before getting behind the wheel. Like Martin and Christine, his drowsy driving was brought on by poor sleep related to a sleep disorder; unlike Martin and Christine, however, Stan dealt with his sleep disorder for years without medical diagnosis or treatment.

Mary, a young woman at the time of her interview, exhibited narcolepsy symptoms, including hypnagogic hallucinations, sleep paralysis, and excessive daytime sleepiness, epitomizing the inevitable forces arrayed against the sleep-disordered driver. The untreated narcoleptic driver, like the idealized norm of the sleepless American, catalyzes the many fears that surround the figure of the drowsy driver: not only might narcoleptics fall asleep suddenly while behind the wheel, but they might also dream that they are still awake and driving. Upon awakening, they might experience momentary sleep paralysis, depriving them of the muscle control needed to drive, even though they are awake and aware of the potential danger of driving while asleep; cataplexy might similarly hamper their ability to drive in times of stress. Mary brings these many factors together and mixes them with the fantasy of being able to continue to drive despite being asleep:

There are days I will not drive, there are certain times I will not drive, or if I'm in the car and realize that I'm having a hard time keeping my eyes open, I'll just go home. Or, if I can't, I'll stop and get a very large coffee and drink as much as I can. It's funny because when I worked—it's not funny, it's actually kind of scary—I would drive on [the highway] for about forty minutes every morning and afternoon, and I just recall a lot of times being in the car on the way home from work or in the morning and feeling like I just needed to shut my eyes, and that's really what it always felt like. For me, what it feels like when I'm having a nap attack, I guess, is it just feels like my brain is almost just being pulled out of the back of my head. That's really the only way I can describe it. And my eyes start, kind of like REM, feeling like they're going blurry, and I just need to shut my eyes for a minute. There were a lot of times I was in the car and I would get that feeling, and I just would think—if it was an open stretch and no other cars were around me, I would think, "I wonder how far I can get with my eyes closed." And it just—that's so dangerous and stupid, but at the time—I never did it, but I wanted to. I would half shut my eyes and think, "I could go straight."

Like an unguided missile, the narcoleptic asleep behind the wheel—or any sleeping driver for that matter—might speed along until a collision occurs or momentum is lost. More often, as these few cases evidence and as laboratory experiments have corroborated, microsleeps occur, lasting only a few seconds, which is long enough to cause serious damage.[17] What Mary brings to the fore are the competing inevitabilities of the drowsy driver, namely, the desire for sleep and the intimacies that compel the sleepy driver to continue driving despite the potential accidents that might occur. What Stan and Martin attempted to allay through their respective use of exercise and stimulants was the symptoms rather than the causes of sleepiness: rather than stopping and taking a nap, they attempted to eradicate their sleepiness temporarily through secondary efforts. Thus, instead of properly addressing the mounting sleep debt they carried, they circumvented it temporarily by stimulating their bodies. Their sleepiness would return, and although awareness of it would lead to further circumventions, eventually the debt would be too great to handle with such temporary treatments. But their intimacies, chemical and otherwise, balanced their desire for sleep, the result being a precarious situation.

One of the problems with Maggie's Law is that the only drivers it identifies as dangerously drowsy are those who have just gone for twenty-four hours or more without sleep. As disordered sleepers evidence, more profound predicaments than a day's worth of sleep deprivation occur among those who drive while drowsy. This is not to argue that disordered sleepers should have no driving rights (as mentioned above, most disordered sleepers are acutely aware of their proclivities for drowsy driving and take the necessary means to ensure they are safe drivers) but rather to expose the central problem of drowsy-driving legislation: it identifies only one symptom and fails to address underlying spatiotemporal demands that lead to drowsy drivers. While sleepwalking can be demonstrated solely through biological factors and the culpability of actions committed while sleeping can be deduced by considering only those factors, sleepiness while driving is difficult to determine unless individuals confess to driving while sleepy. But one might rally evidence from social commitments that overdetermine one's inability to sleep enough or regularly, as well as from expectations about oversleeping and productivity. The latter, however, fail to fully explain the conditions that lead to drowsy driving, so, instead, one must turn to the combination of the intimate and desirous circumstances of the drowsy driver.

Reports of drowsy driving at the turn of the twenty-first century dovetailed with consumers' misuse of prescription sleep-aid pharmaceuticals, which led to the development of a new character type, the "Ambien driver." Although largely indistinguishable from drowsy or other reckless drivers while on the road, Ambien drivers, as well as drivers under the influence of other sleep aids, offer an intimate chemistry of social obligations and pharmaceutical dependencies. Reports from Ambien drivers suggest that they take the drug at bedtime and awaken before its effects are completely worn off. As a result, they are still partly asleep while they go about morning routines, including driving to work. Thus, while they feel awake to themselves, they are clearly inebriated by sleepiness and the drug, often leading to the assessment of these drivers as under the influence of illicit drugs or alcohol.[18] This situation, however, would not arise if individuals were given ample time to sleep and conduct themselves before and after work. Unfortunately, American spatiotemporal orders of everyday life are largely based upon functional sleep deprivation and the regular production of fatigue on a daily and weekly basis, resulting in disorderly sleep for many. Laws tend to base themselves only upon fixing symptoms rather than on the causes of sleepiness and its eruptions into the everyday, however dangerous.

Beyond the Law

If sleepwalkers are one possible exception to the law, then insomniacs are the epitome of those who can vigilantly uphold the law. And no insomniac is clearer evidence of this than Evan Tanner, who, like Edgar Huntly, is a work of fiction and, as such, is able to exist as a cultural ideal unhindered by the realities of human biology. Tanner, while serving as an infantryman in the Korean War, was hit by a piece of shrapnel, which damaged the "sleep center" in his brain. He explains:

> A piece of shrapnel. Nothing seemed damaged—it was just a fleck of shrapnel, actually—so they patched me up and gave me my gun and sent me back into battle. Then I just wasn't sleeping, not at all. I don't know why. They thought it was mental—something like that. The trauma of being wounded. . . . Well, they kept knocking me out with shots, and I would stay out until the shot wore off and then wake up again. They couldn't even induce normal sleep. They decided finally that the sleep center of my brain was destroyed. They're not sure just what the sleep center is or just how it works, but evidently I don't have one any more. So I don't sleep. . . . I rest when I'm tired. Or switch from a mental activity to a physical one, or vice versa.[19]

What results is a frenzy of awareness. Tanner explains, "Once I originally adjusted to going without sleep, I had always contrived to have something to do, someone to talk to, something to read or study. No matter how long one lives, awake or asleep, one can never know all that there is to know."[20] Tanner's adventures span the 1960s and fill eight novels, and he finds himself embroiled in one international intrigue after another, largely because of his enrollment in a secret government agency that accepts him as one of its agents through Tanner's own subterfuge. His excellence as a spy is due to his dedication to using his extra eight hours of wakefulness each day to develop language skills and nurture contacts in maligned social movements throughout the world, such as the Flat Earth Society and the Pan-Hellenic Friendship Society. He explains his productivity to his reader in this way: "When you stop to think about it, eight hours out of twenty-four is a lot of time to waste on nothing more interesting than unconsciousness."[21] In each of the novels, he offers a similar explanation of his lack of desire for sleep and the ensuing need to fill his time. His irresolvable insomnia also makes him an excellent nighttime agent and one who can move through multiple time zones without fear of jetlag: Tanner is the epitome of vigilance.

This vigilance, in every one of his adventures, leads him to great feats. As mentioned above, Tanner's period of activity, save for his final adventure, is during the 1960s, at the height of the Cold War. He is sent on missions to the Near East and eastern Europe on numerous occasions and also to Southeast Asia, Africa, and, in one case, to Montreal, to subvert a Cuban attempt to abduct African Americans for a global proletarian revolution. In each case, the agency he works for expects only that Tanner will foment popular revolt against communist governments or otherwise disrupt the status quo of those who are the enemies of the United States at that time. Because of his lack of desire for sleep, Tanner always exceeds whatever expectations are set for him. The following debriefing with his handler provides a sense of Tanner's preternatural abilities as a subversive agent:

> "Stole one plane, blew up another plane. Helped a top Yugoslav to defect, got translation rights to his book. Slipped an even dozen Latvian gymnasts into the States." He turned to me. "You wouldn't have any extra surprises for me, would you, Tanner?" I lowered my eyes. I looked at my shoes, their heels fitted with rolls of microfilm from Kracow. I thought of the various packets taped to my skin, Milan's book, the Chinese documents that Lajos had smuggled to me from Budapest. I looked at one of the other plush leather chairs and smiled at the sleeping form of Minna, direct descendant of Mindaugas, first and last reigning monarch of a free and independent Lithuania. "Nothing else, I'm afraid," I said. "Well, I'm glad of that. Any more, and I'd have trouble believing it myself."[22]

Tanner's adventures are told in the first person, but the interlocutor who is the recipient of them remains unclear; it is surely not his handler though, because in many of the novels he explains to his interlocutor the discrepancies between what he tells his handler and the facts of the case. This deviation is a critical one, and it might be read as a means through which Tanner attempts to render his otherwise unbelievable stories real to his interlocutor. The chain of events that Tanner routinely sets in motion explains in the real world of the reader why certain events have taken place, including popular revolutions in Colombia and Yugoslavia. As his handler notes in the above passage, Tanner's exploits, as he reports them to both his handler and the reader, are barely credible. Like Edgar Huntly's exploits, they read more like dreams or fantasies than reality. But then maybe the reality of a constant insomniac begins to appear dreamlike.

The real case of Robert Ledru, a Parisian detective at the turn of the twentieth century, is similarly incredible.[23] The story of Ledru's bout of sleepwalking is notorious within sleep science and is often recounted in an apocryphal fashion, many times without any citations and sometimes with wrong dates and locations. According to his biographer, Ledru was widely held to be both deeply corrupt and a fine detective. In 1887, Ledru, a syphilitic for some ten years, was dispatched to Le Havre by his superiors at the Sûreté to assist in the case of a number of sailors who had gone missing. His investigation was quickly interrupted when he was contacted by his supervisors in Paris, who asked him to help the local constabulary in a new murder investigation, which was beyond the locals' expertise. Ledru was introduced to the case: A vacationing Parisian merchant, Andre Monet, was murdered on the beach. He was shot at nearly point blank range. His status as a vacationer traveling alone made it particularly difficult to ascertain who might have attacked the man so far from his home and who would have had any reason for doing so. The bullet was from a common firearm, a German pistol, which only further obscured who might have killed the man. Upon investigating the scene of the murder, Ledru noticed particularly misshapen footprints; specifically, the killer's large toe on his right foot appeared to be missing. Ledru was also missing this toe and had awoken that morning to find his socks inexplicably wet. The facts began to add up: Ledru, in a profound syphilitic stupor, had somnambulated to the beach, apparently with gun in hand, a German pistol, along with the files relating to the case of the missing sailors. The victim either had startled him or had otherwise figured into Ledru's sleepwalking and had unfortunately become the brunt of Ledru's violence. Ledru explained this to the local police and his supervisors in Paris, who agreed that Ledru had no reasons to attack the man, and, since his explanation seemed beyond the ability of science and the law at the time to prove and beyond the ability of common sense to clarify, he was exonerated of the crime. He then was placed in jail for observation. Guards attending to him through the night reported that he would somnambulate in his cell. When they placed a revolver containing blanks under his pillow without his knowing, he reportedly arose from bed the following night, gun in hand, and shot at the attending guards. For fear that he might sleepwalk and kill again, he was committed to a farm in the Parisian countryside, where he spent the remaining fifty years of his life, under guard and attended by nurses.

Why bring this sleepwalker and this insomniac together? Why bring together these two faces of the law, one beyond the law, the other a policeman? In their relationship to one another as ideal embodiments, they show how perverse the law can be, how it depends on an exuberant vigilance, that it is everywhere at once, and yet is corrupted by its unconsciousness, its ambiguity. That is, the law finds its strength in the ambiguity of its application and powers. This is not to argue that the law should be purposely ambiguous, that it should proliferate attempts to capture every possible antisocial action that individuals might take through purposely obtuse conceptions of social behaviors; nor is it to argue that lawmakers should proliferate laws that govern actions through very explicit language, as in the case of Maggie's Law. Both of these legal strategies are inherently problematic, but neither (at least early in the twenty-first century) is a dominant governing strategy in American law. Rather, American legal institutions embody the ambiguity at the heart of the law's power. Regardless of the will of individuals, the law has difficulty with the control of intent: sometimes, despite a driver's best efforts, as in the case of drowsy driving, his or her intentions are subverted by a desire for sleep that cannot be controlled. Just as the biological criteria for the sleepwalking defense required elaboration, so too do the biological exceptions to the law require a fuller elaboration by those in power who see biology as beyond the control of the law and the possible punishments it metes out. Or, if they wish to pursue the prosecution of criminals who would otherwise escape through recourse to biological desires, they may work to recapture the cultural or behavioral bases for crimes, as in the case of drowsy driving.

In this ambiguous formation of the law, the intimate investments of American institutions in the human body become legible. Through the discursive demarcations of human behaviors for which an individual can be held accountable, the unmarked remainders identify those instinctual or animal behaviors, those biological desires, that are beyond the scope of the law and its powers. But in the ambiguous framework that the law provides for thinking about the human body and its capacities, behavior and biology incessantly contaminate each other in their relationship, requiring constant interventions and extrications. This is clear in the case of sleepwalking as a legal defense, where it can be read as a behavioral or intentional action, as in the case of Edgar Huntly or Steven Steinberg, or alternatively where it can be read as a biological problem, an inevitable desire, as in the case of Kenneth Parks. The categories of the natural and the cultural are constantly at play in the legal deployment of the sleepwalking defense and in legislation regarding "drowsy" and "Ambien drivers." In the case of drowsy driving, the nature of sleepiness is viewed through the lens of an individual's ability to choose when to fall asleep. The perceived decision to fall asleep also depends upon the understanding of the act of falling asleep as always being an active decision on the part of the sleeping perpetrator. In making this decision, the momentarily sleeping driver can be held accountable above and beyond what a verified sleepwalker might be held culpable for, because the latter has moved into the realm of the biological while the former remains strictly within the bounds of the social. The desire of the masses—that inevitable biological need to sleep after working too long—is within the realm of the law, whereas the desire of the individual is more dubious. Any reformation of the laws regarding either the use of sleepwalking as a defense or the intentionality of the drowsy driver will invariably shift the boundaries of what can be considered as the biological and the willful. Such mutations in the legal conception of the human invariably also expand and contract what might be legitimately attributed to the causal forces of nature and the arbitrary whims of culture. In this light, the boundaries of the body and its internal and external controls become necessary sites for contesting the realms of the natural and the cultural in American law and the possibilities of sleep.

12

· · · · · · · · · · · · · · · · ·

The Extremes of Sleep
War, Sports, and Science

"Jump Happy" Chutist Takes Leap in Sleep
Atlanta, May 12—First Lt. Gene Padgett of the 501st Parachute
Battalion, Fort Benning, Ga., thinks he must have been "jump
happy." He made a leap last week, he says, and his "chute didn't
open." He reached for the emergency rip-cord, and nothing happened. Fumbling for the rip-cord, he was startled to find he did not
have on his jump suit. "Landing on your feet is the correct way to
jump," he mused. "I landed square on them, but I woke up on the
ground ten feet below the window of the bachelor officers' quarters."
He had been parachute jumping in his sleep.

—*Chicago Daily News,* May 12, 1941

THE U.S. MILITARY HAS HAD A LONG-STANDING SCIENCE FIC-
tional interest in the control of sleep. In the early 2000s, a new
Defense Advanced Research Projects Agency (DARPA) project was
begun, the Continuous Assisted Performance (CAP) project. This project attempted to eliminate the need for sleep through technologies, primarily pharmaceuticals and electrical stimulators. The science of and
experimentation with controlling and eradicating sleep are questionable
and depend on a variety of technoscientific fantasies, which take as their
object a profound reordering of human biology. This particular interest
in sleep is not only a military one but also evident in extreme sports (notably, the Vendée Globe, a round-the-world, individual yacht race) and

science, as in the case of Michel Siffre's experimentations with circadian rhythms while living in a subarctic cavern in the 1960s. Below I address each of these engagements with the extremes of sleep, turning finally to a rare neurological disorder, fatal familial insomnia (FFI), to evidence the limits of these projects. All of these attempts to ascertain sleep's functions and controls, including the CAP project, have tested the limits of the human body and helped to reshape dominant ideas about the nature and capacities of human sleep.

These scientific, sports-based, and military projects are all experimentations with extremes. The extremes bring together science fictions and normative everyday regimes through technics. This is to say that the extremes begin with assumptions about normalcy, extend these assumptions to their ends, and then insinuate the resulting new models of normalcy into the everyday. Through this process, the extremes transform the present into the future, thereby unfolding new possibilities, capacities, and normative regimes. Rather than conceptualizing normalcy as a historically determined project, which simply reproduces itself, researchers have used extremes in a practice that aims to actualize possible futures in contemporary models of the normal. One means through which this operation occurs is science, broadly conceived. But this is science as science fiction, or vice versa. The extremes take as their object ahistorical and acultural conceptions of primordial human nature and relationships between humans and their natural environment. When humans are put into extreme environments, their primordial natures can be revealed; when human nature is pushed to its extremes through technological means, new conceptions of the normal can be elucidated. The extremes generate a world in which new orders make sense of the bodies that they produce, not unlike how science fiction produces new understandings of the everyday. These fantasies of the extremes are not solely about a select few individuals who participate in technoscientific projects, but about all humans, because models of biology and its potentials change along with the invention of new forms of life. In the case of sleep, this invention can be seen in the military attempts to redefine the capacities of the human, in athletes' and scientists' attempts to reconfigure their use of sleep, and in scientific projects that attempt to isolate human biology from society. But these extremes are always defined by and balanced with what human bodies are actually capable of. These human limits, as the history of military science shows, are constantly redefined by new technological developments and new concerns for what to do with human bodies.

The extremes might be approached phenomenologically, by querying the subjectivity that individuals are induced to embody in their experiences with exceptional conditions. However, the extremes are about setting new rules and expectations for bodies: in doing this, they also bring into being new expectations of behaviors. The extremes, rather than providing any functional explanation of themselves, in which they are safely held outside the normal to evidence what constitutes normality, initiate the creeping of new orders, new framings of bodies and behaviors, which slowly emerge as hegemonic conceptions of everyday practices and potentials: science fiction becomes reality. Thus, in the oscillations of the extremes and the normative, a subtle reformulation of both categories occurs: the normal changes on the basis of the extremes, and the extremes are moved to new domains to again produce new norms. The three cases considered here are: the attempts of DARPA to create sleepless soldiers, which highlights the pharmaceutical and technological reordering of sleep; the experimentations with circadian rhythms by French speleologist Michel Siffre in an attempt to ascertain primordial biological capacities outside society; and the training of microsleep patterns of competitive sailors, which capitalizes upon these experiments with human biology and offers a glimpse of one possible new order of sleep and social life. These cases are squared against the realities of sleepless bodies in the form of fatal familial insomnia, which may, at least temporarily, serve as an understanding of where the limits of human bodies lie.

Sleep at War

Like many of the projects undertaken by DARPA scientists, CAP hoped to achieve military success precisely in the potential of the science fictional. DARPA developed what later became the Internet but also spent years researching "remote sensing," or clairvoyance, and a host of other dubious possibilities. These fantasies mixed the limits of human bodies with natural and technological means in an effort to redraw the limits of embodiment and the capacities of the human. The CAP project mixed a pragmatic goal, the eradication of sleep deprivation's ill effects, with research agendas that were on the controversial frontiers of science. As discussed in a declassified presentation from one of the CAP researchers:

> Studies have shown that sleep deprivation results in poor judgment and lowers physical performance. The ability to maintain focus and respond to the world around you is lessened, and these are precisely the attributes that a warfighter needs most. The need for sleep is a significant limiting factor for the warfighter, and, by removing it, an immediate advantage is gained in the form of a force multiplier. Continuous operations are possible, and the tempo for those operations is accelerated. Imagine if we could remove the need to sleep for periods up to one week without a reduction in our ability to process sensory input, make decisions, and respond to the external environment with focus and intention.[1]

The rhetorical shift from the customary *soldier* to *warfighter* is a perverse one, denoting the rationale for eradicating sleep deprivation's ill effects in the soldiers of the future: there will no longer be a need for the auxiliary social aspects of soldiering, only the brute force of war itself. "Imagine," Joseph Bielitzki asks his audience, "if we could remove the need for sleep," leaving its possibilities ominously open; sleep's control will be measured in the "force multiplier" that sleep's superfluity confers on future warfighters. What this force multiplier might be is nebulous, as are the side effects of such a radical transformation in human biology.

The goals of the CAP project were described by Bielitzki in the following terms, which left aside the more violent aspects of sleepless soldiering:

> Preventing changes seen in the brain that are caused by sleep deprivation; expanding or optimizing available memory space within the brain to extend performance; rapidly reversing adverse changes in the brain caused by sleep deprivation; and developing problem solving circuits within the brain that are sleep resistant.[2]

What any of these goals might actually lead to and how they might be brought about were left undetailed—what, precisely, might a "problem solving circuit" in the brain be, other than a quasi-cyborg imaginary? Although this language evokes a solider who is able to apprise his or her environment fully and respond to it effectively, what it conceals about the sleepless brain is the violence that it is expected to administer through the body of the warfighter. The supposition underpinning these goals recognizes that sleepiness and the need for sleep are primarily seated in the brain and that fatigue, aching muscles, or injuries sustained by the body might be overridden by a more capable nervous system. This model of sleep as behavioral and subject to choice rather than biological and subject to physiological demands is a popular one in research regarding sleep's eradication; in most experiments with the possibilities of human sleep in extreme situations, various ways of engaging the mind have been deployed in an attempt to keep the body vigilant, but the brain itself is supplemented with technologies and techniques of management.

As is to be expected, the development of these supersoldiers relies in part on existing technologies but supplements them with contemporary scientific techniques, including genetic analysis, "novel" pharmaceuticals, and means of affecting the brain directly with electricity and magnetism to produce higher cognitive functioning while sleep deprived. These technoscientific techniques rely on the perceived differences within human bodies but also supplement these natural bodies with prostheses that extend them toward the future in novel ways. As described in a declassified DARPA document:

> The Continuous Assisted Performance (CAP) program is developing a wide range of technical approaches to extend the capabilities of soldiers to perform their duties for up to seven days without sleep. Unlike existing brain stimulators, such as caffeine or amphetamines, CAP will develop techniques that maintain cognitive function during long periods of sleep deprivation. . . . [The] portfolio of efforts . . . include: (i) magnetic brain stimulation; (ii) understanding individual differences in resistance to sleep deprivation; (iii) effects of exercise and diet on resistance to sleep deprivation; and (iv) the discovery of novel pharmacologic approaches. . . . the program is identifying technology to successfully reverse the effects of sleep deprivation on executive function. In addition, the program is expanding approaches that enhance neurogenesis as a protection against the effects of sleep deprivation.[3]

Some of these approaches are hardly science fictional, as in the case of ascertaining the effects of exercise and diet and their relation to sleep; one can easily imagine such knowledge leaking out into American society more generally to produce new alertness-promoting food fads and sleep-friendly diets. But is there room in American society for "magnetic brain stimulation" salons? How these developments and reconfigurations of the human body might trickle into American society more broadly is entirely unforeseen and wholly avoided by the DARPA scientists associated with the project, at least in print. The knowledge of the successful development of such sleepless bodies may result in a Lamarckian shift in culture, if not biology: if the possibility of such a radical transformation is confirmed, then it might become the new foundational understanding of human biology, the extremes being the new normative regime. If it can be done, why not for every American?

Science in the twentieth century stood in its own way in developing sleepless soldiers. The basic mechanisms of sleep remained undiscovered by scientists, and in this ignorance, the development of sleepless bodies was only a fanciful possibility. In the language of CAP's director, Amy Kruse, this was put into the relational rhetoric of being in the "far-term":

> For the far-term, we are studying the mechanisms of sleep—in creatures from flies to humans—in an effort to uncover the critical components, neural substrates, and modulators of sleep. . . . On a microlevel, we now know that specific genes can regulate sleep and wakefulness needs, and these genetic sequences will be used to identify candidate biomolecules that might eliminate the deleterious effects of extended sleep loss.[4]

The genetic understanding of sleep, its causes and obviation, brings together the evolutionary past with the high-technology, science fictional near future. Such future-talk brings together time and biology in ways that firmly root the future in the biological, not in the cultural or social. In this version of the future, biology and knowledge of biology are ascendant, and social life and cultural expectations will be reshaped by intimate knowledge of how sleep might be controlled. The danger of such a prospect, or rather of this line of thinking, is that contemporary science is "far" from being able to explain basic functions of sleep and its abatement, and fantasies of such, if taken increasingly as reality, might initiate new orders that restructure everyday life in ways increasingly antagonistic to sleep. Such scientific fantasies may initiate a twenty-four-hour society that relies upon and expects fully vigilant bodies. But turn-of-the-century science was unable to make bodies that could maintain a twenty-four-hour existence for a long enough term to make sleeplessness a tenable fantasy, let alone a new natural order. But experiments with fundamental human nature are nothing new: consider Michel Siffre's solo adventure into the caves of the French Alps in order to elucidate the basic functions of circadian rhythms.

The Primordial Nature of Sleep, Again

In 1962, Michel Siffre descended into a subarctic cave in the French Alps. He had previously led expeditions into caves throughout Asia, northern Africa, and mainland Europe. For Siffre, this expedition had speleological and geological importance as well as potentially profound impacts on the study of human circadian rhythms. His account of the experiment is presented in *Beyond Time,*[5] which comprises his philosophical musings on the nature of the experiment, his edited diaries as recorded during the two months that he resided in the cave, and ample evidence of his unfailing enthusiasm for spelunking. In Siffre's words, his intent was to prove the social and environmental impacts on reckoning time, because he perceived the patterns of human behavior as primarily conditioned rather than natural: "Down in the cavern I would isolate my personal life rhythm from all cosmic and social references. I remembered how excited I had got in philosophy class over the studies made by Pavlov. My intuition was that men and animals alike are conditioned by the regular alternance of day and night. Would I, living underground, rediscover the original life rhythm of man?"[6] What Siffre sought in his descent into the subarctic cave was biology devoid of society, the very foundations of animal being, which by his presumption had been obfuscated through centuries of accruing civilization and normative spatiotemporal regimes.

But the site of the experiment was fraught with environmental hazards, and Siffre was in questionable condition for the experimental protocol. Siffre experienced both incidental problems (notably, some anxiety about the experiment and attendant insomnia previous to his descent into the cave) and ongoing gastrointestinal issues. He disclosed the following: "I spent a restless night turning and turning in my sleeping bag, thinking about what was to happen next day, the day so long awaited, now so near and frightening. . . . Toward dawn, utterly worn out, I sank into a deep sleep. The camp was already bubbling with activity when I woke up."[7] After awakening and on his way to the cave opening, through which he was to scale down 130 feet for the purposes of the experiment, Siffre was struck with another health concern: "Then—oh, the weakness of human flesh!—during this rapid and tiring climb on the frontier ridge, I suddenly fell victim to a violent attack of amoebic dysentery";[8] Siffre's experiences of the weakness of human flesh were only just beginning, because his time in the cave would lead to serious health problems for him. Siffre began his experiment with not only a sleep debt but also intermittent dysentery—hardly an ideal specimen for such experimentation. Both of these factors became evident in his recounting of his time in the cave, although he failed to reconcile both of these concerns with his expectations of what the experiment would demand and evidence.

To wit, Siffre's average sleep length extended slightly over eight hours in every twenty-four, in contrast with his life in Paris, where he claimed his nightly sleep was either six or seven hours. While in the cave, his appetite virtually disappeared: he ate only one meal per "day" and often reported eating not out of a perceived hunger but because he felt he should. His infrequent eating was compounded by his poor cooking skills and his failure to equip himself properly for the experiment; having a shortage of funds for the expedition, he had bought provisions last and with little forethought. Moreover, in gauging the date, he rationed much of his most appealing food for the second month of his stay underground, and when the experiment ended, half of his food stores were still waiting to be eaten. And, finally, the cave was often flooded, soaking his food, his tent, and his clothes in near-freezing water, which ruined much of the stored food and kept Siffre in bed for long periods. This malnourished and sleep-deprived man—although his sleep may have been more representative of today's contemporary average than he knew at the time—was the "everyman" who would provide a basis for understanding the naturalness of human circadian rhythms and their conditioning, largely through his own dedication to the experiment, but also with the unflagging assistance of an entire team of people who acted as observers, travel companions, and support staff. The issues of a low sample size, his health, and the experimental conditions were never raised or at least were not presented in *Beyond Time;* Siffre seemed to have convinced his colleagues that the experiment was sound and that its results would provide a meaningful glimpse of human biology in a social vacuum.

Not only was Siffre seemingly oblivious to his own biological failings and the shortcomings of the experimental conditions, but he also carried ideas about the naturalness of time and its arrangements with him into the cave. Siffre was intensely aware of the subjective and objective differences in time, which constituted a central part of his experimental protocol: each day, upon awakening, again at meal time, and finally before bedtime, Siffre would call his handlers to inform them of what time he thought it was. These phone calls were intended as one-way communication, and Siffre was precluded from having conversations about anything other than his perception of the time with the people on the other end of the phone. His subjective sense of time was then recorded alongside objectively measured time. Siffre explained:

> [I] was to telephone the surface when I awoke, when I was ready to eat a meal, and when I was about to go to sleep; in this way, my hazarded time of day could be checked against the actual time which those on guard must never tell me. I kept a chart of my subjective "time" and they kept a chart of the actual time. Thus we were able to calculate the length of my periods of repose and activity in the course of twenty-four hours, and the quantitative extent to which I lost my sense of time. . . . My only way of estimating time was by means of my physiological functions, and these functions in man have been conditioned since the beginning of his existence on earth by the regular alternation of night and day, so we thought.[9]

Siffre immediately began to vary from objective time, at first off by about four hours and at the end of the experiment off by two weeks. This variance was attributed to a phase advance and was compounded by his sense of what time and day it should be. At three points in the course of his diary, Siffre attempted to deduce the day, averaging his number of awakenings with what day he thought it must be. His surprise when his handlers told him the experiment was to come to a close was due to his fuzzy mathematics, which, despite his knowing that his subjective sense of the time was very likely wrong, held that the end date was nearly a month away. Siffre's established perceptions of time and their value shaped the very structure of the experiment; he explained, "[I] was not to be brought to the surface for any reason whatsoever before the passage of a month's time below. My experiment would be meaningful only if it lasted that long."[10] Why a socially normative length of time such as a month? This choice seems to have had nothing to do with the perceived rhythms of biology, the production of proper experimental conditions, but only with Siffre's perception of time, which quickly changed, and no length of time underground would have put Siffre's perception of time back in synch with the objective time measured aboveground. Whether this had to do with the experimental conditions or with Siffre himself is debatable.

While in the cave and properly severed from human contact, Siffre endured serious strains on his mental health. This may have been due to his removal from society, the conditions of the cave, his neurochemical predispositions, a combination of these factors, or others that Siffre failed to account for and detail. His diary entries swing wildly between manic self-congratulations for his bravery and scientific cunning and woeful bouts of depression, and Siffre admits to having edited the journal-as-presented for content, having removed what he deemed redundant. What remained was a fairly explicit account of manic depression, of a man alone in a subarctic cave.

> Willfully I cut myself off from time, that regulator of man's occupations; I wanted to find out how a man behaves when so cut off, and had reverted to a primitive animalistic way of life in which everything was subordinated to my natural needs. I was no longer a slave, either to men and their social habits or to the effect of the rotation of the earth on its axis. I told myself, "I am free!" But was I in fact free?[11]

Aside from the complication of Siffre's existentialism, the cave was a particularly difficult environment for him to inhabit, with its lack of natural light, its constant flooding, and its near-freezing temperatures. Only because of his equipment could Siffre maintain his physical well-being in the cave, and even this perseverance was besieged by his depression. He recounted, "Rising was a harsh test of courage; the sleeping bag was the only place where I found relative comfort. By habit an early riser, I struggled against this torpor. I now believe that if I had given way to it I would not be alive to tell the story."[12] Siffre ended his account with an appeal to "young aspirants" not to overvalue their equipment or their physical healthiness but rather to emphasize what he accepted as an innately human trait, the power of the mind: "Will power plays a part as important as, if not more than, the careful choice of equipment and a rigorous course in physical training. I would tell young aspirants that, armed with this weapon, you can do a great deal; you can do anything."[13] In light of Siffre's own mental health concerns while underground, this may seem strange, but only in identifying something that he perceived to be exclusively human could Siffre maintain scientific credibility: if the same experiment could be conducted with rats, dogs, or any other animal species, what is the point of putting a man in a subarctic cave for two months?

What did Siffre's experiment actually show, after he had spent two months in a subarctic cave, exhibiting signs of depression, malnutrition, and sleep deprivation? Siffre was convinced that it showed the adaptability of the human body to extreme environments, which he summarized this way: "It showed how a man, but no means an athlete, was able to adapt himself to an environment differing greatly from the conditions existing on the surface of the earth."[14] Moreover, it evidenced for Siffre the innate superiority of the human species within the animal kingdom, because it indicated that humanity alone could overcome such adverse environments. He wrote, "There are situations in which animals perish but in which a man can survive, thanks to his mental powers and his astonishing ability to adapt to new environments."[15] It is telling that in his appraisal of his survival within the cave, Siffre credits his mental powers, not the various equipment or the food stores that he brought with him into the cave. His crediting of his mental powers despite their failure at times might seem paradoxical, but because his experiment was to test the human frame and its biological ability to adapt to adverse situations, he had to overlook the issue of technology. Siffre not only obscured the importance of technology and the other conveniences he brought into the cave; he also wholly took them for granted as part and parcel of humanity and the human experience of the world.

Again, though, what was Siffre able to evidence during his two months alone? He described what doctors ascertained upon his removal from the cave: "Nothing was fundamentally wrong with me. I was merely suffering the effects of extreme fatigue, and these minor ailments would vanish with a return to normal life."[16] The utility of this knowledge is dubious, although it was promising enough to lead NASA in the 1970s to conduct an experiment with Siffre at its heart, which involved keeping him in a climate-controlled cave for six months in an attempt to ascertain how the effects of long-term estrangement from environmental and social rhythms might impact the biological lives of astronauts.[17] To this day, Siffre's "experiments" are more akin to extreme sports than they are to rigorous scientific testing: in both experiments it was Siffre alone against his sense of time and the environment, which hardly allows for reproducible scientific conditions; in both experiments, it was not so much "man versus nature" as "man and his technologies versus nature." What this could evidence was necessarily limited, but the scientific and technological control of sleep was often pursued in other contexts, again testing the limits of the human body and its mastery of itself and the environment.

The Sport of Sleep

The control of sleep has played a central part in the Vendée Globe, a one-person yacht race from the coast of France around Antarctica and back to France. The race can take over three months to complete, with many racers taking over four months to cross the finish line. The stretches between France and the Southern Ocean are generally problem-free, although never wholly so, and around Antarctica the sailors need to be constantly vigilant, because the dangers of strong winds, icebergs, and other environmental hazards are imminent. Sleeping for long periods under these circumstances can seriously endanger both the sailor and his or her ship; unlike Siffre, underground with handlers only 130 feet above him, sailors in the Vendée Globe have no ready escape hatch, and their death has come to be an expected occurrence during the course of the race. Under these conditions, experimentation with sleep has become a necessity, because sailors experience chronic sleep deprivation, sometimes going for months without a full day's ration of sleep. As might be expected, caffeine has played a central dietary role, but sleep's manipulation has been of primary importance, as detailed by sleep researcher and sailor Claudio Stampi:

> Today, the limiting factor is no longer technology—virtually all competitors sail on extremely fast and state-of-the-art racing machines—but the human element. Races are won by solo sailors who, pitting themselves against nature's elements for months at a time, are capable of wisely administering their own resources of stamina, skill, organization, self-discipline and determination. The key to success in these

great human adventures and athletic contests is proper management of sleep and rest. For these solo sailors, the temptation to reduce sleep to dramatically low levels is constantly present at any time of day or night in order to continuously optimize boat performance and speed, to survey tactics of competitors and study meteorological reports, and to avoid collisions with ships or with icebergs in the Southern Oceans.[18]

As Siffre implied with his experiment, Stampi suggests that technology can largely be taken for granted; the human element is of greatest concern, which is presumably better managed by some competitors than others. Take, for example, Pete Goss, who competed in the 1996 Vendée Globe, and his account of the race. Goss evidenced the role that sleep and sleeplessness played in his experience of the race and, much like Siffre before him, struggled with his biological predispositions in an environment radically removed from society. More than Siffre with his health and emotional problems, Goss serves as a biological everyman, fit but not overly so, lonely at sea but not existential in his plight, and despite the many tragedies that befell him, Goss was dedicated to remaining healthy while at sea. The extremes of his situation, with his constant lack of sleep, might parallel in effect if not in content the future imagined by military science, wherein sleep is wholly circumvented. Before we turn to Goss, however, the role that sleep's manipulation took in the race and the science that underpins sleep are worth further discussion, because they expose assumptions about biology and human nature widely held in contemporary science.

Stampi's research on "ultrashort" sleep was the state of the art at the turn of the twenty-first century and the result of a long-standing personal interest in understanding how the human body can survive with only minimal sleep. His approach within sleep science was heterodox but represented the place of napping within contemporary sleep medicine, namely, that napping might be employed in situations where vigilance is needed and wherein other more conventional chemical or technological means of assuring such levels of awareness are restricted or already maximized. Thus, Stampi's research protocol involved highly regulated periods of sleep and wakefulness, shortened overall daily sleep, and objective and subjective attempts to measure the performance of ultrashort nappers. Stampi's most compelling experiment involved monitoring the sleep and daily activities of a lone subject who was subjected to forty-eight days of sleeping only three hours per day in an attempt to model Leonardo da Vinci's mythical sleep patterns of fifteen minutes every four hours.[19] The test subject became more difficult to awaken as the test went on but once awake performed well on all tests that were assigned to him. It was this model of sleep that Stampi exported to the Vendée Globe, and long-distance boat racing more generally, although the model remained rather resistant to incorporation into American everyday life.

Stampi was famously consulted for managing the sleep of Ellen MacArthur, the youngest participant in the Vendée Globe and only the second woman to compete in the race: "What Ellen is doing is finding the best compromise between her need to sleep and her need to be awake all the time. The best compromise appears to be like cats and dogs and most animals, which is to break up sleep into short naps."[20] Other sailors also often drew upon his expertise, and it was incorporated into articles in the Vendée Globe online magazine, which offered insight into how the various sailors prepared for their race and managed their sleep. The following passage, which, despite construing these naps as something primordial and natural for all of the animal kingdom (much like Stampi's mention of "cats and dogs and most animals" above), interprets the ability to nap as something almost athletic, requiring training to achieve competitive napping skills and maximum efficiency:

> It's up to each participant to discover his own needs and his own rhythm and to learn how best to manage them. It's a very special skill, which takes a long time to acquire and you have to develop it constantly in order to maintain it. Hoping to win these ocean racing sportsmen must also become top level sleepers.
>
> Why these cycles of sleep? Our sleep pattern is based on the genetic make-up from prehistoric man, who could not remain asleep for 8 hours without endangering his life. He woke up at the end of each cycle, had a look around before going back to sleep for another cycle. With the absence of predators, modern man no longer needs to wake up so often. So we sleep in one stretch, bringing together all the cycles in one. Sleeping in single periods is an ancient ancestral predisposition, which the ocean racers, facing danger, must find again.[21]

Nappers, it seems, are exceptional in their ability to tap into their ancient ancestral predisposition, effectively overcoming their enculturation and expectations of normal sleep and everyday life. In the successive article on the sleeping habits of the Vendée Globe racers, the temporal limits of such manipulation with sleep are discussed, although somewhat paradoxically, because the author of the article claims that, on the one hand, the technique of such strategic napping can be used for only a limited period of time before adverse side effects begin to plague the sailors and that, on the other, many sailors find it difficult to adapt to "normal" sleep once ashore.[22] To be a competitive sailor in the Vendée Globe has meant also being a competitive sleeper, and acquiring the ability to alter one's sleep schedule efficiently and at will has set many racers apart from the otherwise untrained, sleep-deprived sailors who have attempted to maintain normal sleep patterns while at sea.

Pete Goss, by his own admission, was a competitive sleeper. He described his ability to sleep at sea in the following way, nearly an ideal embodiment of Stampi's ultrashort napper:

> I love my eight hours' sleep when ashore, but can reduce it to about four in every twenty-four when competing; it's a necessity if you want results. I break my four hours into twenty-minute catnaps. It's at least ten days before my system settles into it. Until then I feel old and ache all over.[23]

Throughout *Close to the Wind,* Goss makes casual references to his general lack of sleep, his desperate need for caffeine, the carelessness of his actions while sleep deprived, and the uncomfortable conditions for sleep, such as the need to strap himself into bed while circumnavigating the Antarctic to avoid serious injuries incurred when thrown from bed after hitting large waves or other obstacles. Goss was an ideal sleeper, able to adjust to the demands of his racing, but also able to love his eight hours of sleep when ashore, and his ability to shift between modes of sleep may have brought him to the following realization about the role of sleep in everyday life in England. Identifying the gentle tyrant of sleep in the ordering of social life, he wrote:

> I had an hour's deep sleep and bounced out of bed refreshed. In normal society, everything is geared to that eight-hour sleep. Transport, shopping, radio, television—it goes on and on, ruling your life, confining you like a straitjacket. At sea my pattern varies with the weather and although I feel tired for much of the time I am not debilitated.

Physically I feel lean and mean. My pain barrier rises considerably and a knock that would have hurt like buggery ashore is shrugged off as if it were nothing. There is the odd day when my limbs feel heavy, my eyes are gritty and I catch myself gazing into space in a kind of exhausted trance.[24]

Such sleepers as Goss—and all humans may share the same biological potentials that he has—are the most likely road to scientifically identifying means for adjusting sleep patterns in extreme conditions. But because this adjustment relies simply upon the reordering of sleep and not upon its eradication through pharmaceuticals or other technologies, it is a strategy that is available to any society, any army, and fails to confer on any one group or individual a tactical advantage, unless their enemies are too confined by the "straitjacket" of consolidated sleep. Goss's experimentation with his own sleep was a response to the perceived conditions that he would confront at sea; his experiment, like Stampi's, might lead to a radical reorienting of everyday life for those who are in pursuit of temporal advantages and are willing to flirt with the possible side effects of sleep deprivation and desynchronization from normative models of everyday life. But this is only one possibility of reorganizing human sleep; another lies in the genetic makeup of those diagnosed with fatal familial insomnia, a chronic form of sleeplessness.

Sleep at the Threshold

Chronic insomnia of a sort that maintains alertness and decision making—despite the U.S. military's attempts to produce it alchemically—has been a rare occurrence and is known in only one form, fatal familial insomnia (FFI). By all accounts, FFI is an incredibly rare prion disease and is genetically transmitted through about thirty family lines, located primarily in Europe and the United States. As such, it has rarely been clinically observed in detail, and no cure for it has been identified; most often, patients are admitted for care only once the disease has firmly taken hold of their social and biological lives, and their deterioration commences quickly thereafter. Those eventually diagnosed with FFI descend into increasing disorder, often resulting in a compromised immune system and eventually death. By examining what was known about FFI early in the twenty-first century, we apprehend a cautionary tale of human limits and confront the biological fantasy of a sleepless human body, its capacities and its ends.

The onset of FFI usually occurs between age thirty-six and sixty-two, and from the time of diagnosis, individuals can live anywhere from eight months to six years. Individuals generally experience a severe decrease in their sleeping times through the night and, as a result, lapse into fugue states throughout the day; concomitantly, they develop parasomnias, such as sleepwalking and dream enactment. Throughout this intensification of symptoms, individuals are resistant to traditional sleep aids. Pasquale Montagna and his laboratory colleagues, who diagnosed the first cases of FFI, describe this descent into apparent madness: "Worsening of sleep and autonomic disturbances is associated with the onset of peculiar oneiric behaviours, whereby patients, especially if left to themselves, fall into a hallucinatory state and display motor gestures related to the content of a dream; these symptoms can be mistaken for psychotic signs."[25] FFI is hardly the sleeplessness that is desired or imagined by military scientists, athletes, or individuals beleaguered by their daily social obligations, but something more perverse; just as narcoleptics embody sleep, these genetic insomniacs embody wakefulness in ways not as idealized by military researchers but as supported by the human body. There is, it appears, a limit to how sleeplessness can be embodied, and FFI may be its representation. FFI is a slow torture, not a productive, alert means of being in the world; it is a radical disruption of basic human biology, and one that leads, inevitably, to an estrangement from everyday life, despite the afflicted's ability to be present at all times. In spite of this counterexample of sleeplessness's desirability and the examples of sleep's potential control at the hands of trained nappers, the pursuit of sleep's eradication remains central to military science and a palpable American fantasy.

I intend no romanticization of the sleeping body or of human biology. What I hope the preceding material evidences are other ways of being sleepers, both fantastic and actual, in an effort to show how disparate the ideal forms of sleeplessness are from actual sleepless bodies, how the extremes produce and rely upon untenable but seductive fictions of human biology. As Jonathan Moreno describes, the need to change the lowest common denominator of the military—the human body—is a persistent fetish, and new manipulations of human bodies may lead to an entirely new kind of "arms race":

The military wants to juice up personnel's brains because the human being is the weakest instrument of warfare. Although for centuries astonishing and terrifying advances have been made in the technology of conflict, soldiers are basically the same. They must eat, sleep, discern friend from foe, heal when wounded, and so forth. The first state (or nonstate) actor to build superior fighters will make an enormous leap in the arms race.[26]

This "arms race," if successful, could lead to new "social races," new configurations of the social based upon altered biologies, whether they inaugurate alert, sleepless bodies or societies of oneric zombies, both extreme forms of life. In either case, the decision about the coming of these new communities may be made not by those whom it will impact the most but rather by the scientific stewards of humanity's biological futures. In concluding this chapter, I offer the penultimate word to David Dinges, noted sleep researcher with a long-term research agenda pursuing the extremes of human sleep, which has been supported in part by NASA: "Now is the time to have an open and frank discussion on how far we will go as a culture. What are our priorities? How regularly do we want to manipulate our brain chemistry? What are the limits?"[27] There are no limits to experimentation with human sleep; there will always be the extremes, and they will be subject to exploration, manipulation, and contestation. And human life and the social order will continue to be shaped and reconceived on the basis of these extremes and their actualizations.

Conclusion

······················

The Futures of Sleep

ARLY IN THE TWENTY-FIRST CENTURY, SLEEP CAPTURES AMERI-
cans' imaginations, with stories often published by *Time, Newsweek,*
and the *New York Times Magazine* and aired by *Dateline NBC.* This focus
is the result of sustained interest on the part of the nonprofit National
Sleep Foundation, medical professionals, pharmaceutical companies,
and various segments of American society who, each in particular ways,
have attempted to make sleep disorders more apparent to the American
public. In doing this, they have also attempted to make the social more
biological in its foundations and to transform the American public into
a sleeping public as well as an alert one. They have worked to make the
slumbering masses something real, a standard that guides our everyday
lives. Often embedded in these popular discussions of sleep are references to a conjectured end of sleep, the dream of a future in which sleep
will be a social luxury and not a biological necessity. Or, rather, where
sleep and wakefulness can be controlled through simple pharmaceutical
means, with sleep-inducing Lunesta for dinner and alertness-promoting
Provigil for breakfast, with wakefulness and sleepiness managed as desires rather than acting as impulses that control the social lives of sleepers. This model of society supplants normal patterns of human sleep with
pathological or fantastic forms; it replaces what sleep Americans do get
with the patterns emblematized by narcoleptics and supersoldiers. In both
cases, normal human sleep is replaced with the science fictional or the
scientific as fictional.

In anticipation of this future, in 1996, at the very moment when interest in sleep was intensifying as a result of the Z-drugs, *Wired* published a brief article titled "The Future of Sleep," in which the article's author asked a group of sleep professionals, including a medical clinician, a medical researcher, and two experts on lucid dreaming, about their predictions for the future of sleep. Drawing on the knowledge of those assembled, *Wired* asked about the specific future of treatment for insomnia, sleep apnea, and snoring, as well as technology to induce dreaming; no consensus was reached among the experts for the possible future of these concerns. What did achieve consensus was the prediction for the possible total elimination of sleep, which all agreed was "unlikely." Now in the early decades of the 2000s, looking back at this *Wired* article as a historical artifact, we still have no "real cure for snoring and apnea," which are different conditions, though both are treated as chronic, and the latter requires nightly prosthetics in the form of CPAP and BiPAP machines; electronic dream inducers are nonexistent; the very idea of an "effective nonaddictive sleeping pill" is debatable, especially as pharmaceutical companies have moved away from the language of addiction to that of "habit formation"; and, moreover, Americans are still sleeping, evidencing the "unlikely" elimination of sleep.[1] But the very idea of the end of human sleep is a persistent and popular one, and its possibility acts as a fantasy of scientific progress—of what science could bring into being—for the practice of sleep science and the American public, creating a subtle pressure on cultural conceptions of what sleep is, how it evolved, and what it can become.

In this chapter, I put this recurrent fantasy into dialogue with the lived realities of disordered sleepers. What this allows me to do is to formulate a social possibility, what I call *multibiologism,* a cultural and medical acceptance of nonpathological variation within species, which recognizes both society and biology as mutable within limits. What I am interested in here is the embodied limits of the elimination of sleep— the defiance of bodies to particular technoscientific futures—as they are produced in the discursive practices of medical practitioners and in scientific literature and as they are lived by individuals. I do not intend any romanticism of human biology as it resists these futures; rather, my appeal to material limits of bodies grows out of my sympathy with scientists and with disordered sleepers. My claims are based in the clinic, not the laboratory or the science fictional; rather than seeing the clinic as solely the site of disciplining and controlling bodies, I want to pull from it a bioethical model that is more ethical than bioethics is currently conceived and practiced. If anything, then, my recourse to human biology is antiromantic, steeped in an awareness of how inevitable sleep is, not solely as a physiological phenomenon, but also as a multifaceted desire. The future of sleepless humans is a fantasy; if humans were ever able to stop sleeping, such an event would index a deeper transformation in human biology, and society, that would bring us squarely into the realm of the posthuman. But until that future is realized or fundamentally disavowed, there will always be a perverse and fantastic horizon for sleepers: maybe human biology as it is currently known will come to an end, and we will all suddenly have more time in our days. Most likely this future will not take place, but until it does or does not, human bodies will be subject to the pharmacological and technoscientific pursuit of this implausible posthuman transformation.

In the context of the practice of sleep medicine early in the twenty-first century, the entanglements of American everyday life, sleep, and the science fiction of sleep's eradication provide a window into understanding the conflicting conceptions of the body in contemporary medicine, its relationship to time and conjectured futures, and the desires it produces and is produced by. These entanglements also open a window into the role of the very concept of everyday life and the ways that patients' "having a life" influence the means and ends of medicine, rendering the practice of medicine as both present-oriented and concerned with producing livable futures for disordered sleepers, as opposed to transforming the basis of our species. In the following, I begin by tracing the understandings of human sleep forward, following its genealogy, briefly, from the work of Nathaniel Kleitman through his successors over the twentieth century. I then focus on the case of a patient who, as understood by attending physicians, could not stop sleeping and on the way sleeping both structured and destroyed her everyday life. This is followed by an interruption from a disordered sleeper, who narrates his entrenchment in social and biological situations that fundamentally shaped his experience of both. These cases help to evidence how sleep's role in American social life, despite the recurrent fantasy of sleep's eradication, is so entrenched that its removal would produce more disorder than sleep disorders themselves do. By way of conclusion,

I offer up the idea of multibiologism and its place in the contemporary practice of bioethics and social life generally. It is the near inevitability of the future of sleep (the unlikely future of its eradication) and its confused nature that produce the rhythm and idea of American everyday life early in this century and sustain dominant conceptions of the body and its limits in American society, and particularly in American sleep science. One way out of these recurrent fantasies of sleep's eradication is through multibiologism, accepting human biological difference, not as determining, not as destiny, but as a foundation for reconceptualizing society and its realization of human desire's manifold potentials.

Conceptualizing Sleep, Again

Nathaniel Kleitman, in his magnum opus, *Sleep and Wakefulness,* reviewed historical explanations of sleep and identified them as following two trends: the humoral and the neurological. Humoral models extend back to Aristotle (at least) and assume that something—a substance, a feeling—in the body builds up to a degree that it needs to be alleviated, and only through sleep can it be efficiently done. These models held sway through the nineteenth century and continue to inform some medical thinking. Neurological models, which began to be developed only in the latter half of the nineteenth century, posit that there is some "center" in the brain that controls sleep and wakefulness. From this center, sleep becomes a "global" phenomenon, synchronizing the rest of the brain and body and producing sleep. It was this model that Kleitman championed throughout the twentieth century and, in one form or another, has been popular into the early 2000s, when it has begun to be supplanted by a "local" model of sleep.

This local model, forwarded by psychiatrist Carlos Schenck and neurologist Mark Mahowald, two leaders in the field of sleep science at the turn of the twenty-first century, who were introduced in chapter 4, posits that sleep is not based in any center but rather emanates from a variety of sources that eventually induce full sleep. Mahowald and Schenck came to this theory through years of studying parasomnia behavior; old models of human sleep could not account for the behaviors of sleepwalkers, sleep-eaters, and sexsomniacs. Nor could they account for Ambien zombies: people technically sleeping but still able to eat breakfast, get dressed, and drive to work. And the understanding of sleep as fragmentary and unconscious allows for individuals' unwillful enactment of behaviors while they are technically asleep. This local model of sleep is revolutionary, in that it challenges two trends that have plagued conceptions of sleep throughout history, namely, that sleep and its control are a matter of willpower and that sleep in its present form can be explained through narratives of human evolution, which often accept human sleep as an adaptation to environmental forces. Taken together, the old assumptions about sleep effectively posit normal human sleep to be inevitable and malleable, but only by those sufficiently well-bred or fit. These assumptions of flexibility and inevitability underlie American expectations of sleep, especially the possibility of its eradication; to get away from these precepts, Americans need to adopt a different model of sleep and human biology altogether, one related to the local model of sleep and its underlying assumptions about sleep as unwillful and variably expressed, within and among individuals.

The eradication of sleep as a function of willpower is best captured in an experiment conducted by Kleitman in the 1930s, an experiment that has often been referred to by other sleep researchers. Kleitman and his team observed a set of graduate students attempting to stay up for four days straight, for ninety-six hours of continuous activity. The first twenty-four hours passed by with relative ease, with the participants complaining of sleepiness when reaching the early morning of the second day. After this short period of sleepiness, the participants were able to stay awake without much effort until late the following night. Only through physical activity were they able to stay up through the critical hours between 3 and 7 A.M. The night of the third day was the same, and individuals who exercised through the night were able to stay alert. When the participants reached day four, Kleitman decided to call an end to the experiment, because he felt the results would persist, supporting the claim that exercise could overcome sleepiness. The conclusion Kleitman argued was:

> In the operation of the multiple feedback circuits, wakefulness capacity is limited by muscular endurance. . . . Subjects could maintain wakefulness as long as they were able and willing to maintain muscular activity. Even a well-rested person may have difficulty in

remaining awake, if, in addition to the removal of, or decrease in, stimulation through other sense organs, he allows his skeletal musculature to relax. Conversely, tense muscles may be responsible for "insomnia."[2]

This was later followed up by William Dement, who in 1965 observed an attempt to break the *Guinness Book of World Records* achievement for the longest period of continuous wakefulness by a San Diego teenager named Randy Gardner. Gardner managed to stay awake for eleven days (264 hours), beating the record by 4 hours. How he managed to do it was primarily by playing basketball through the night, often against Dement or his assistant. Looking backward, Dement now recognizes that Gardner was constantly taking micronaps, falling asleep momentarily and reawakening, unperceived by either onlookers or himself. Given this, no one will be able to stay awake by modern standards for the length of time recorded for Gardner. Nonetheless, contemporary sleep researchers appeal to Gardner's breaking of the Guinness record and the results of Kleitman's experiment to argue that through physical fitness and willpower, individuals can defy the biology of sleep. This is best represented in Peretz Lavie, who conducts sleep reduction research with the Israeli military and who argues:

> People who are absolutely determined are able to continue without sleep for several days, especially when they have the support of constant monitoring and excitatory stimuli. As the only way to fight off sleep is through energetic physical activity during the crucial period of the early hours of the morning, physical fitness is very important. Fit people are simply better equipped to withstand prolonged periods without sleep.[3]

The assumptions that Lavie and Kleitman share are fundamental to orthodox understandings of sleep: determination, fitness, endurance, and being able and willing are essential to control wakefulness and sleep. Although they depend fundamentally upon some physiological foundation, all of these qualities can be attained and developed through physical and mental practice. They also depend upon an acceptance of "neo-Darwinian" human biology that posits flexibility as a cultural and social response to millennia of evolution;[4] a fit nonsleeper can realize his or her own biological future through dedicated effort.

The assumption underlying what Kleitman refers to as "primitive sleep," or sleep out of necessity, and evolutionary just-so stories is that humans have developed consolidated sleep as an adaptive response to environmental cues. Since the social and natural environments have radically changed, so too might human sleep. This is best epitomized in the rationalization of sleep posited by Michel Jouvet, who, among his peers, is foremost in his conceptualization of the evolutionary ecology of human sleep. Jouvet writes:

> The waking state, however irregular, is a necessary condition for the realization of most vital homeostatic regulation. We need only think of the search for food and its absorption, the search for a sexual partner and mating, the hatching of eggs or suckling of young, the defense of one's territory, and many others. It could be that each of these functions is prepared or regulated during sleep. However, sleep depends of necessity on the external environment.[5]

But then how do scientists explain contemporary human sleep patterns and the possibility of overcoming them? How can sleep be eradicated if it is necessarily part of the human species? Dement has the answer, and it lies in the possibilities of contemporary American science—and science fiction. He argues, following Jouvet:

> All organisms must meet an energy budget. They cannot expend more energy in growth and activity than they bring in by consuming calories. Each animal best hunts or grazes for these calories under different light conditions. . . . In order to meet the animal's energy budget, there is a strong pressure to curtail activity when the light and the hunting are not good. When the time of day is wrong for calorie gathering and may even make the organism more vulnerable to becoming someone else's dinner, it's better to lie low and be quiet than risk becoming food. This survival strategy is programmed directly into the genes in the form of a drive to be periodically inactive. . . . For humans in countries where there is a steady food supply (and, in some places 24-hour supermarkets), this isn't a factor. If saving calories is sleep's primary purpose, why do the well-fed citizens of postindustrial societies still need sleep? In part because it's only been a second or two of evolutionary time that calorie conservation hasn't been extremely important—and in developing or famine-prone countries, it's still essential. From this perspective, we may someday shed the inherited urge of the sleep drive or eliminate it through genetic engineering.[6]

This is to say that the next evolutionary shift in human sleep may be in adapting to both the presence of twenty-four-hour supermarkets and the availability of easy calories. And for those in famine-prone countries, they now have the possibility of a genetic arms race ahead of them, as well-fed citizens of postindustrial countries breed themselves into a sleepless future, or at least the elite among them do. But all this overlooks the problems of the American economy and eating patterns: the twenty-four-hour day has receded over the past decade, and the presence of easy calories has seemed to lead more quickly to diabetes than a reduction in sleep need. But if the twenty-four-hour day is realized, in all its science fictional potential, and if Americans adapt to a high-efficiency diet, then Americans may be evolving into a posthuman future that aligns with Nancy Kress's novel *Beggars in Spain,* which sees the genetic engineering of sleep's eradication as necessarily a problem of class and integral to the further development of American capitalism.

Kress's Beggars series provides a bridge between the assumptions of twentieth-century sleep science and dominant American cultural expectations of society and its relationship to the possibilities for sleep's future.[7] Kress's novels of this series take place in a future in which a limited number of select, wealthy humans have eradicated the need for sleep, and hypercapitalism is the result: the "sleepless," drawing on their parents' already established fortunes, quickly dominate the global economy. In their biological transformation the sleepless also lose the ability to dream and so become highly rational, calculating individuals—embodiments of a fantastic capitalism. While the rest of the world sleeps, the sleepless conspire in their elite community for their individual and collective benefit. As one might expect, the sleepers (those "normal" people who still sleep) quickly begin to resent the lack of biological restraint the sleepless benefit from, and the procedures that produce sleeplessness quickly fall into disuse as a result of these social tensions—the poor resent the technology, and the rich fear the wrath of the poor. The sleepless are exiled to an asteroid that orbits Earth, and from there they continue to control the planet's future, economic and otherwise. This is a necessarily reductive summary of Kress's work, which also includes global conspiracies against humanity and new breeds of the "supersleepless" to contend with, but what is clearly evidenced in these novels is the commonly held understanding of the link between sleep and humanity, of how essential sleep is in producing communities, in maintaining an order and rhythm to society, and in determining what has become increasingly accepted as the American iteration of the Protestant ethic and its embodiments.

What the sleepless endure is a Protestant ethic run amok, in which the early bird never even has to go to bed. What they sacrifice for this power is their creativity, but who needs it anyway? As American capitalism and its agents are popularly imagined, the sleepless become coldly calculating, able to abide by the rules of the economy but unable to envision new ones, working only toward their own gain and damning those unlike them. Kress, however improbable her conjectured future is, is correct in diagnosing this problem of capital, or rather American problems with capitalism: by the start of the second decade of the twenty-first century it is becoming difficult to imagine everyday life and its economic basis in any other form; capitalism's future contaminates and structures the present in pervasive and widely accepted ways. The sleep disorders discussed below, and throughout this book, are disorders in the sense that they fail to abide by the economic rationalities that Americans have imposed on themselves in their social formations but that also make sense of twenty-first-century forms of life as a result of the conjectured inevitability of the market. What results is the proliferation of sleep disorders, produced by the imposition of emergent spatiotemporal orders of a particularly rigid social order that takes the Protestant ethic too seriously, which has produced the chemical supplements to enforce that order and the cultural expectations to legitimate their use.

Hugo Gernsback, the father of American science fiction and an engineer by training, proposed a more modest future in which humans would be able to capitalize on their sleeping through their ability to educate themselves subliminally with prerecorded educational programs.[8] Such a conceit assumes that the brain is as passive as the rest of the body appears to be during sleep. What has become known since Gernsback's 1921 conjecture is that the brain is anything but passive while the body is at rest; sleep is one of the brain's most active periods, and sleeping as a process is appropriately construed as demanding more of the brain than any other individual part of the body. The problem disordered sleepers most often face is the alignment of their biological demands and their social lives, something even Gernsback seemed aware of; for Gernsback though, social desire could be routed into sleep through technological means. As imaginative as this approach might be, it fails to be imaginative enough; it fails to reimagine the taken-for-granted social, biological, and cultural structures that form everyday life. Might Americans, in tandem with sleep researchers and clinicians, not rethink the demands of the workday, of the school and family, and, in so doing, reimagine human biology as involving not monolithic norms but individual forms of life, as human biologies? Otherwise Americans may be doomed to a future of proliferating sleep disorders, amphetamine breakfasts, and sedatives for dinner, a way of life that increasingly emphasizes chemical supplements to make sense of human social and biological desires. Instead, sleep's variabilities might be articulated to establish manifold understandings of everyday life as flexible and open to experimentation. The groundwork for this reconceptualization of sleep and social life has already been laid in sleep medicine, but it requires inverting the assumptions of sleep's medicalization.

Inverting Sleep

Sleep is always biological and cultural. It is always social and individual, economic, obligatory, and necessary. Like other basic human physiological functions, it is mired in and constitutive of these various domains: the economic, the social, the cultural, and biological. And because of this richness or messiness, we need to develop bioethical models that can accept and address the intimate problem that sleep often is. With these claims in mind, I turn to the sleep clinic where I conducted my fieldwork, bringing together two cases of problematic sleep and showing how they might invert the medicalization of sleep.

Consider the case of a woman who was a "long sleeper," but because of the rhythm of her social obligations, sleep insinuated itself wherever it could. Dr. Palmer, a research fellow who came to the clinic in my second year of fieldwork, presented the case of this young woman, a single mother of five children, who reported sleeping seventeen hours each day. She reported that she went to bed every night after *The Jerry Springer Show* (earlier if it was a bad episode), which at the time ended at 11 P.M., and would awaken at 8 A.M. to prepare breakfast for her children and to see them off to school. She then slept again from 10 A.M. until 12 noon, at which point she awoke to watch her soap operas, which she struggled to stay awake through, then slept between 2 P.M. and 6 P.M., and upon awakening would make dinner for her family. If these times were accurate, she slept no less than fifteen hours on a daily basis. She explained that as a child she would stay inside during recess and sleep while other kids went out to play and that she had two sisters who were the same way. Dr. Richards, one of the senior neurologists, explained that the woman was convincing in her narrative and symptoms and that there was no evidence of mental disorder (other than her watching *Jerry Springer*). Because of the regularity of her sleep, the patient was treated as hypersomnulant, which required placing her on alertness-promoting drugs, the same as those prescribed to narcoleptics. Follow-ups on the part of the staff at the clinic turned up nothing about her further sleep problems; the staff assumed that she was noncompliant and that she continued to sleep as she had for decades. Society, in this case, was already in line with her needs as a sleeper, or the patient had already acclimated her biological demands to her social desires.

At the time I interviewed Ryan, a middle-aged white man, in 2006, he had been diagnosed with just about everything a disordered sleeper could be diagnosed with: narcolepsy, REM behavior disorder, obstructive sleep apnea, shift work sleep disorder, and a vague circadian rhythm disorder. Only at age forty had he decided something might be physiologically wrong, despite exhibiting symptoms since childhood. And only at forty-eight did he finally seek diagnosis. At the time of our interview, in his midfifties and inching toward retirement, Ryan had some control of his sleep through a mixture of pharmaceuticals, CPAP technology, and social arrangements of his working time. "I work a twelve-hour shift," he told me, "from six at night until six in the morning, or from six in the morning until six at night." He was employed by a large power company on the East Coast, working to maintain the integrity of the power grid of a large metropolitan area. His workday consisted of his sitting in front of a control console for hours at a time, with little change in activity or object of focus: dull work, but within his unionized labor force a sought-after position, since it involved no handling of any electrical equipment and hence was not life endangering. Because of his host of sleep disorders—and workplace problems narrated presently—he took Provigil, an alertness-promoting drug, at work. He went on to explain not only his work situation but also how it rendered his sleep disorderly:

The longest one shift goes is four days, and then I shift to the nights. And I can have one day off in-between or eight days off in-between. . . . And then there's one week when you have to work relief, where you have to work four hours in the morning, then twelve hours that night, and twelve hours the next day, so my biggest problem is "when do I take my medication." If I have to skip it, then I'm more of a zombie. . . . I took a letter from my neurologist that said that I need to take a midday nap on each shift, and they sent me home for three weeks without pay while they figured out what to do. They brought me back and said, "If you take a nap, you're fired." And this is a company with twelve thousand employees. And then I took a letter in that said that if I continue to work without napping, I could endanger myself or others—and with that one they sent me home for three months. . . . I was on "crisis suspension," so I got paid for that one. . . . My personal feeling is that they don't want anyone to have any kind of personal accommodation or anything, because it will open up a can of worms. [My sleepiness was] troublesome when I was a kid, but the older I get, the harder it gets.

"How do you cope with it?" I asked. "Napping, and working an eight-hour shift. I think napping works. But my employer treats napping as a personal choice, so that means it's a conduct issue. That's what they believe right now." This belief is the result of policies based, if only tacitly, upon the evolutionary and neo-Darwinian assumptions embedded in contemporary sleep science and American spatiotemporal regimes of work, school, recreation, and family life. If the desire to take a nap were viewed not as a "personal choice" or a "conduct issue" but instead as the result of a complex intimacy of biology, culture, and society resulting in a particular desire for sleep, Ryan and the other patients I have discussed herein might not be interpreted as disorderly as their sleep presents them. One route toward this bioethical stance is an acceptance of physiological variation within the human species: multibiologism.

Variability without Difference: Multibiologism

Since the industrial revolution, Americans have become invested in a form of social organization that limits varieties of sleep. Moreover, this limiting organization also forces the medicalization of particular normal sleeping patterns, rendering them pathological. The social arrangement of time—the workday, the school day, recreation, and family life—has no absolute basis in nature. Instead, any perceived or claimed alignment is simply an ideological just-so story that naturalizes hegemonic understandings of time and its relationship to human biology. By all accounts, up until the mid-nineteenth century, Americans slept differently: they shared beds with family members beyond the marital couple, slept with babies through childhood, and fragmented their sleep, napping during the day but also waking throughout the night and staying up for hours. Insomnia as Americans know it today came into being only with the invention of the workday, one that expected continuous performance and vigilance. Work, then, became intimately tied to caffeine and sugar and the colonial projects that provided those products to imperial centers. This is the legacy of modern sleep, an artifact of these colonial and industrial projects, consolidated not through the tendency of human nature but through social organization and cultural expectations of normalcy and productivity. Changing the workday, school day, and cultural understandings of family life and recreation may be an impossible task, but it is one that might be pursued through a bioethics that takes as its aim the production and facilitation of other forms of life.

Bioethics, as it has been practiced in the United States since the 1960s, has largely involved biomorals. Bioethical practitioners (physicians, philosophers, theologians, lawyers) have primarily focused on the cultural problems associated with new technologies, especially as they are applied to an increasingly medicalized American populace. Is it right or wrong to terminate a pregnancy given knowledge gained through amniocentesis? How long should one wait to remove the life support of a family member in a persistent vegetative state? Is cloning a technology that should be used for therapeutic or research purposes? How should stem cells be used, and which stem cells are viable for medical deployment? These are questions of right and wrong, of good and evil, of transcendence; they are also cultural questions, resolvable only contingently. New technologies, or refinements of extant technologies, force these questions to be asked again: there is no end to the biomoral. If bioethics is to become actually ethical instead of focusing on transcendental concerns of good and evil, it must take as its concern the potentiality of life itself. Instead of adjudicating right and wrong, bioethics needs to address how social organization makes life possible and what forms of life and desire it allows and what forms it renders difficult. This is to say that the focus of bioethics should not be technologies and their use, or decisions regarding individual lives, but rather the entanglements of human biology, social obligations, and cultural expectations and desires. Bioethics should take as its aim not good acts but the promotion of varieties of forms of life.

I come to this bioethical position not through the usual route (what I take to be the usual route of bioethics, namely, through Descartes and Kant) but through Spinoza and Deleuze. Instead of pushing the question of transcendence, of good and evil, Spinoza and Deleuze push the ethical toward immanence and potentiality. They suggest that what makes life's many expressions possible is of the greatest good. The ethical thought Deleuze gives us, through his readings of Spinoza, is an ethics of life itself. Deleuze's concern is primarily with what he refers to as desire, although not any form of conscious desire. Instead, his concern is with the unconscious production of desire, desires that are the products of and producers of the worlds within which individuals are situated. Desire, in Deleuze's formulation, is that which ties the biological and individual to the social and environmental. As a force, it is irreducible. But it is variable, and this is where potentiality and possibility come into play: the broader the social possibilities, the more vectors desire has to follow. When society has only one normal or natural model of sleep, desire is necessarily limited; when sleep can be more flexibly arranged, desire is necessarily more variable as well. If sleep is accepted as actually more diverse than it is in the American consolidated model, then the resulting bioethical problem is multiplying the possibilities of sleep and society as they are currently lived.

In the clinic, a multibiologism is already operating in the relationships among patients, doctors, and treatments. Some patients respond positively to a particular drug, and others develop side effects; in the latter case, doctors quickly place patients on new drug regimes, attempting to alleviate negative symptoms associated with treatment. The plethora of antidepressants and sleep drugs involves more than simply a variety of consumer choices to be made; these drugs often have minor chemical differences that have profound impacts on individuals and their social relationships. The appropriate and effective use of these different chemicals depends upon the recognition that the human species is variable and that even minor variations sometimes respond in profound ways to slight changes in chemicals. Moreover, physicians often allow for a diversity of expressions of sleeping as normal sleep, recognizing that the recommended eight hours of sleep each night is an average, with most humans sleeping between seven and nine hours nightly; when patients vary from the norm, they, more often than their physicians, are the ones to instigate a modification in their sleeping patterns, requesting a therapeutic intervention. In the case of many sleep disorders, the average becomes a normative ideal: individuals and American society itself have become invested in maintaining an average that proliferates abnormalcy when it isn't achieved. Delayed and advanced sleep phases, insomnia, and short and long sleeping periods are problems only when individuals recognize them as disorders, and at this point doctors intervene to adjust sleeping times in accordance with patients' desires. The recurrent fantasies of sleep science and science fiction can be tempered by clinical experience and the lives of disordered sleepers. Since nearly all Americans are disordered sleepers at some points in their lives (with occasional insomnia, a delayed or advanced sleep phase, night terrors), the social arrangement of sleep might be brought into accord with the recognition that everyone is affected by the fleeting, and sometimes inevitable, demands that these disorders cause. Normal sleep is always pathological sleep, or at least potentially so. But the pathological is also within the variations of the human species.

An anthropology that ignores human biological variation cedes its powers to the sociobiologists and evolutionary psychologists. It could be otherwise, though: anthropology at its core is about understanding human variation, and human biology offers both a "universal" foundation for analysis and a basis for comparison across cultures. But this is to argue that human biology is not determining but, instead, "local."[9] Humans sleep, but how and when we sleep, what we consider restful and disordered, with whom we sleep, and in what arrangement—these are all variable. Moreover, they are variations that may help anthropologists understand how culture and biology can interact. This also suggests that the repressive hypothesis, in all its variations, needs to be disposed of. There is no repression, only integration and acknowledgment in the intimate assemblages that make up human desire: our social obligations, cultural expectations, and biological exuberances. Understanding the deep history of American sleep patterns—their interaction with family life, recreation, work, and school, their medicalization, their intimate ties to economic forms and forces—is a first step in acknowledging that human sleep makes and is made beyond nature; no causal force exists, but rhythms that make and remake society and all of its expectations do.

The Therapeutic Horizon

At the annual sleep medicine meetings in Salt Lake City in 2006, there was no more spectacular display than Rozerem's. The meetings are large, attracting about six thousand participants: doctors, researchers, lab technicians, the occasional sleep activist or patient, and manufacturers. Among the many manufacturers at the meetings, the pharmaceutical companies were the most ostentatious. Lunesta, for the second year running, was represented by a large merry-go-round video installation; visitors to the booth waited their turn to lie in an ergonomic chair aboard the merry-go-round, where they reclined for ten minutes to watch an informational video about the drug. Having free-standing monitors would have failed to attract people, but the conceit of the merry-go-round seduced people into docility for the video's length. Upon leaving, the visitors were gifted with a free copy of a recent book, *From Angels to Neurons,* by sleep pioneer J. Allan Hobson, who, in his late eighties at this point, happily scrawled out autographs to suitably adoring scientist–fans. Provigil, meanwhile, had a series of multimedia installations on large, touch-manipulated flat screens that offered visitors an opportunity to test their knowledge about the human nervous system, sleep, and alerting mechanisms through interaction with beautifully rendered 3-D images. Other, smaller companies offset the need for spectacle with swag—flash drives, mugs, tee shirts, notebooks—which attendees happily lined up for.

But Rozerem had installed a thirty-foot-tall zero, a reference to their new advertising campaign, which referred to "zero evidence of abuse or dependence in clinical studies." The zero stood on a short base and extended some fifteen to twenty feet above that. Located at the front of the corporate displays in the convention center, it was commanding. But the icing on the cake was this: Rozerem had its female drug representatives alternate in taking turns feigning sleep in the base of the zero. Every couple of hours a shift change would take place, and one of the drug reps would emerge from the nearby women's bathroom in her pajamas, ascend the base of the zero, pull a white blanket over herself, and go to sleep—or pretend to.

On some level, this was a very canny advertising campaign. Groups of attendees were often standing around, trying to figure out if the woman sleeping in the zero was real or not; she was just far enough away to make it difficult to tell. And representatives on the ground took that as an opportunity to promote Rozerem to distracted attendees, who were often more interested in the sleeping woman than they were the effects of the drug. Despite this and a later advertising campaign that starred a beaver and Abraham Lincoln, by all accounts—financial and clinical— Rozerem has failed to achieve blockbuster status and limps behind many of the off-patent sleep aids in number of prescriptions. Whereas Ambien continues to offer new versions of its basic, good-selling drug, Rozerem has floundered. And this is because of the drug's effects and how they fail to accord with American expectations of sleep—the very reason that there's no evidence of dependence or abuse. Although effective, Rozerem fails to align individuals with the spatiotemporal ordering of American everyday life in quite the way they desire.

Rozerem is a drug with a short half-life, meaning that its effects quickly wear off, in this case, after about four hours. So for those who awake in the middle of the night, the drug will have worn off by then, and you will need to take another pill; Ambien, on the other hand, ensures that you will sleep through the night uninterrupted—and maybe through your morning routine, your drive to work or school, and the first couple of hours after leaving the house. Which would you choose for your un-consolidated sleep? One might rightly argue that Rozerem is the perfect drug for sleep onset insomniacs—whenever they need to sleep, a drug is ready to help put them to sleep. If they wake in the middle of the night, Rozerem will get them through the rest of the night without overdosing them. But this ignores that what many American sleepers are concerned with is *waking up in the middle of the night.* The giant zero at the 2006 sleep meetings now feels like an ironic portent of sales to come.

American everyday life relies upon pharmaceuticals that enable and reinforce the roles of our many institutions in everyday life, from schools to work, from family life to medical care, from the economy to recreation. Moreover, everyday life as we know it is produced by pharmaceuticals that engender particular effects and rhythms valued as American. The success or failure of a drug, especially sleep drugs, has everything to do with cultural assumptions about nature and productivity, and despite the historical and cross-cultural evidence to the contrary, American sleepers tend to believe in the veracity of consolidated sleep through the night, so as to consolidate work through the day. This is especially evident in the drugs Americans take to be good, orderly sleepers. This limitation of possible therapies and plausible solutions to disorder defines our therapeutic horizon, one defined not so much by limitless potential as by the values embodied in the pharmaceuticals produced and the ways they affect bodies. Our emergent pharmaceutical futures are defined by the spatiotemporal forms that we value and that accord with our cultural assumptions of life, productivity, and normalcy.

Over the past two centuries of industrial and postindustrial American capitalism, the production of laboring bodies has moved from the stimulation of bodies through sugar and caffeine to their stimulation through pharmaceutical chemicals, such as modafinil (or Provigil); the chemicals have changed, but their intended usage has remained constant. These chemicals are intended to produce laboring bodies, alert, capable, and punctual, in accordance with their scientific management, and, in so doing, to reify expectations of nature and normalcy. Integralism has guided these collaborative developments of medicine, market, bodies, and spatiotemporal orders, folding individual desires into the desires of the masses.

Our desires are malleable, as the shift from biphasic to consolidated sleep evidences, as do our emergent desires for novel forms of stimulation and sedation. What serves as a barrier to the expansion of our desires is the limits of human capacity. We may never produce a soldier capable of being active for 196 continuous hours; nor may we elucidate the powers that compel the behaviors of the wayward sleeper. Embedded in the many attempts to reconceive American sleep—the Take Back Your Time movement, Metronaps, the workplace napping consultancies, and the revision of school start times—are also new desires. The failure of these many attempts has less to do with their efficacy than it has to do with the ways they accord with dominant expectations of normalcy and American desires for sleep. In the reshaping of the contours of our desires, our biologies, societies, natures, and cultures mutate as well. Similarly, our markets and medicines, our labor and productivity, transform. How, when, where, and why we sleep are always political; they have the potential to remake our desires and our spatiotemporal orders.

Recently, a great deal of attention has been paid to the molecular focus within the life sciences, particularly as genes have taken on explanatory powers for disease susceptibility, behaviors, and mental and physical capacities. But I would like to change the emphasis of our critique of the effects of this molecular focus within everyday life: attending to the ways that Americans have relied upon chemical supplements for the formation of laboring, productive, alert bodies helps to evidence the long and complicated history that desire has had with the chemicals available through American medicine. Americans may be meeting contemporary demands with new chemical compounds, but this approach only recasts old desires in new capsulated forms: we have always been chemical subjects, but our complex chemical histories are yet largely unwritten. What wages these chemical histories demand from our conceptions of normalcy, nature, and diversity are as yet similarly dormant. Our histories and everyday lives are more discreet, more invisible than we allow, obscured through the patina of everyday life's seemingly inevitable spatiotemporal rhythms and orders. Therapies that formulate new orders may be troubled, but they offer one route away from the machines of integralism; the failures of these therapies index the social powers they have and how unsettling those might be. Our futures are as yet unwritten as well, and although they emerge from our present conditions and concerns, their horizons defy prediction.

ACKNOWLEDGMENTS

· · · · · · · · · · · · · · · · ·

THE DEBTS INCURRED IN WRITING MY DISSERTATION WERE many. Those incurred in its transformation from that state into its present form were much more intimate.

My dissertation committee, Bruce Braun, Jean Langford, and Thomas Wolfe, helped me move from dissertation to the first draft of the book; Jean, especially, continued to push me in my thinking about medicine and its social and somatic effects. Faculty in the medical school at the University of Minnesota helped me understand sleep and its richness. I am especially indebted to Mark Mahowald, Carlos Schenck, and Michel Cramer-Bornemann, who convinced me sleep is worth thinking about. Financial support for this research was provided through generous grants from the University of Minnesota and the National Science Foundation's Science, Technology, and Society directorate. The Andrew W. Mellon Foundation provided generous funding through the University of Minnesota's Institute for Advanced Study for the completion of the book manuscript—and for my beginning work on its sequel.

My dissertation adviser, Karen-Sue Taussig, single-handedly convinced me to continue my education and training as an anthropologist. She has long been supportive and thoughtful in her provocations, and this book would be less without the benefit of her guidance. Along with Jon Kahn and Emma Kahn Taussig, Karen-Sue gives me a family away from home whenever I find myself in Minnesota. Or Amsterdam.

Colleagues at Wayne State University, especially Jacalyn Harden, offered me excellent feedback on my numerous talks there. I also benefited from audience comments and conversations at the University of Chicago,

Acknowledgments

Rensselaer Polytechnic Institute, Stanford University, the University of Vienna, and Rice University. I have accumulated many friends and colleagues over the years of giving talks for anthropologists and others, and Davin Heckman, Sam Collins, Seth Messinger, Tiffany Romain, Stephanie Lloyd, and Gretchen Bakke deserve special mention for their continued friendship.

My colleagues at the University of California, Santa Cruz, have been generous with their time and insight. Don Brenneis and Lisa Rofel have been dedicated and supportive mentors; both read this book in a much earlier state. Lisa deserves special acclaim for reading it again—and for pushing me to clarify my thought even further. Danilyn Rutherford and Nancy Chen both engaged with the text in generative ways, and I am indebted to their readings. This book has taken the shape it has as a result of my formative early career years at the University of California, Santa Cruz; if I were somewhere else, it may not be what it is.

I am especially grateful to my editor, Jason Weidemann, who sought me out after hearing about my work; he saw a future for the book that I didn't. My thanks to Jason and his fellow editors at the University of Minnesota Press, who dare to promote thought and are wonderful hosts. Stacy Leigh Pigg was tasked with producing two reviews of the manuscript, both of which were appreciated, as was a third review from an anonymous reader.

My deepest thanks go to my partner, Katherine Martineau, who has been living with this book, in one form or another, for more years than we should recount. She has been patient as I stumble through explanations, and she has been a pragmatic and demanding editor. Without her, my sentences would be longer. She has endured my many frustrations, sleepless nights, and textual anxieties. I couldn't ask for a better bed partner. Although he will never read this, I also acknowledge our dog, Turtle, who ensured that I wasn't always at my desk and that, when I sleep, my feet are rarely cold. And thanks to Felix, our son, who is (generally) an amazing sleeper: if he were to sleep in any other way, this book would not be in your hands today.

Notes

·················

Preface

1. The "Midwest Sleep Disorders Center" is a pseudonym, as are all the names of individuals encountered during my fieldwork and interview respondents. Demographic information is used in the text as provided by respondents.

2. Allan Rechtschaffen and B. M. Bergmann, "Sleep Deprivation in the Rat: An Update of the 1989 Paper," *Sleep* 25, no. 1 (2002).

Introduction

1. Thomas A. Edison, *The Diary and Sundry Observations of Thomas Alva Edison* (New York: Greenwood Press, 1948), 58.

2. There is an extensive body of literature on the production of normative regimes, ranging across disciplinary interests from psychoanalysis to the history of medicine, from post-structural philosophy to disability studies, including, but not limited to, Georges Canguilhem, *The Normal and the Pathological*, trans. Carolyn R. Fawcett (New York: Zone Books, 1991 [1966]); Lennard Davis, *Enforcing Normalcy: Disability, Deafness, and the Body* (New York: Verso, 1995); Lennard Davis, *Bending over Backwards: Essays on Disability and the Body* (New York: New York University Press, 2002); Gilles Deleuze and Félix Guattari, *Anti-Oedipus*, trans. Robert Hurley, Mark Seem, and Helen R. Lane, *Capitalism and Schizophrenia*, vol. 1 (Minneapolis: University of Minnesota, 1983 [1972]); Michel Foucault, *Madness and Civilization: A History of Insanity in the Age of Reason*, trans. Richard Howard (New York: Vintage, 1988 [1961]); Michel Foucault, *The Will to Knowledge*, trans. Robert Hurley, vol. 1 of *The History of Sexuality* (New York: Vintage, 1990 [1976]); Michel Foucault, *The Use of Pleasure*, trans. Robert Hurley, vol. 2 of *The History of Sexuality* (New York: Vintage, 1990 [1984]); Michel Foucault, *The Care of the Self,* trans. Robert Hurley, vol. 3 of *The History of Sexuality* (New York: Vintage, 1988 [1984]); Michel Foucault,

Discipline and Punish: The Birth of the Prison, trans. Alan Sheridan (New York: Vintage, 1995 [1975]); Michel Foucault, "Technologies of the Self," trans. Robert Hurley, in *Ethics: Subjectivity and Truth,* ed. Paul Rabinow, vol. 1 of *Essential Works of Michel Foucault, 1954–1984* (New York: New Press, 1998 [1982]); and Sigmund Freud, *Three Essays on the Theory of Sexuality,* trans. James Strachey (New York: Basic Books, 2000 [1905]).

3. For a comparative example, see A. Roger Ekirch, "Sleep We Have Lost: Pre-Industrial Slumber in the British Isles," *American Historical Review* (2001). I discuss the American case in chapter 2 and in Matthew Wolf-Meyer, "The Nature of Sleep," *Comparative Studies of Society and History* 53, no. 4 (2011).

4. For a discussion of changes in American work time, see David Roediger and Philip Foner, *Our Own Time: A History of American Labor and the Working Day* (New York: Verso, 1989).

5. For discussions of objectivity in scientific practice, see Lorraine Daston and Peter Galison, *Objectivity* (New York: Zone Books, 2007); Donna Haraway, *Modest_Witness@Second_Millennium.FemaleMan©_Meets_OncoMouse™: Feminism and Technoscience* (New York: Routledge, 1997); and Bruno Latour and Steve Woolgar, *Laboratory Life: The Construction of Scientific Facts* (Princeton, N.J.: Princeton University Press, 1986 [1979]).

6. For a discussion of how the normal is known through the pathological, see Canguilhem, *The Normal and the Pathological.*

7. As I discuss presently, my understanding of everyday rhythms is derived from the work of Henri Lefebvre. See, for example, Henri Lefebvre, *Everyday Life in the Modern World,* trans. Sacha Rabinovitch (New Brunswick, N.J.: Transaction Publishers, 2002 [1971]); and Henri Lefebvre, *Rhythmanalysis: Space, Time and Everyday Life,* trans. Gerald Moore and Stuart Elden (New York: Continuum, 2004).

8. See, for example, the works of Phillippe Bourgois and Jeffrey Schonberg, *Righteous Dopefiend* (Berkeley: University of California Press, 2009); Paul Farmer, *Infections and Inequalities: The Modern Plagues* (Berkeley: University of California Press, 1999); Paul Farmer, *Pathologies of Power: Health, Human Rights, and the New War on the Poor* (Berkeley: University of California Press, 2003); and Nancy Scheper-Hughes, *Death without Weeping: The Violence of Everyday Life in Brazil* (Berkeley: University of California Press, 1992).

9. Foremost among those influenced by Foucault are Margaret Lock, *Encounters with Aging: Mythologies of Menopause in Japan and North America* (Berkeley: University of California Press, 1993); Emily Martin, *The Woman in the Body: A Cultural Analysis of Reproduction* (Boston: Beacon Press, 1992 [1987]); Emily Martin, *Flexible Bodies: Tracking Immunity in American Culture—from the Days of Polio to the Age of Aids* (Boston: Beacon, 1994); Rayna Rapp, *Testing Women,*

Testing the Fetus: A Social History of Amniocentesis in America (New York: Routledge, 1999); Lorna A. Rhodes, *Emptying Beds: The Work of an Emergency Psychiatric Unit* (Berkeley: University of California Press, 1991); and Lorna A. Rhodes, *Total Confinement: Madness and Reason in the Maximum Security Prison* (Berkeley: University of California Press, 2004).

10. Representative contributions to medical anthropology from the science studies perspective are diverse, including Sarah Lochlann Jain, *Injury: The Politics of Product Design and Safety Law in the United States* (Princeton, N.J.: Princeton University Press, 2006); Erin Koch, "Beyond Suspicion: Evidence, (Un)Certainty, and Tuberculosis in Georgian Prisons," *American Ethnologist* 33, no. 1 (2006); Michael Montoya, "Bioethnic Conscription: Genes, Race, and Mexicana/o Ethnicity in Diabetes Research," *Cultural Anthropology* 22, no. 1 (2007); and Karen-Sue Taussig, *Ordinary Genomes: Normalizing the Future through Genetic Research and Practice* (Durham, N.C.: Duke University Press, 2009).

11. Lefebvre's discussion of the everyday is extensive and is the basis of much of his work. Two representative and succinct texts are Henri Lefebvre, "The Everyday and Everydayness," *Yale French Studies* 73 (1987); and Lefebvre, *Everyday Life in the Modern World.*

12. For a discussion of the cultural context and development of the calendar, see Eviatar Zerubavel, *Hidden Rhythms: Schedules and Calendars in Social Life* (Berkeley: University of California Press, 1985 [1981]).

13. Deleuze discusses habit throughout his corpus but in a sustained fashion in Gilles Deleuze, *Difference and Repetition,* trans. Paul Patton (New York: Columbia University Press, 1994 [1968]). The discussion of "bodies without organs" occurs in Gilles Deleuze and Félix Guattari, *A Thousand Plateaus,* trans. Brian Massumi, vol. 2: *Capitalism and Schizophrenia* (Minneapolis: University of Minnesota, 1987 [1980]).

14. Stiegler's discussion of technics is ongoing, but begins in Bernard Stiegler, *Technics and Time,* vol. 1: *The Fault of Epimetheus,* trans. Richard Beardsworth and George Collins (Stanford, Calif.: Stanford University Press, 1994).

15. One genealogy of the emergent as the processual derives from Alfred North Whitehead, discussed in Steven Shaviro, *Without Criteria: Kant, Whitehead, Deleuze, and Aesthetics* (Cambridge, Mass.: MIT Press, 2009); and in Alfred North Whitehead, *Process and Reality* (New York: Free Press, 1985 [1978]).

16. For a discussion of how anthropological writing produces alternate temporal frames for its objects, see Johannes Fabian, *Time and the Other: How Anthropology Makes Its Object* (New York: Columbia University Press, 1983).

17. The social science literature on expertise is extensive. For a review, see E. Summerson Carr, "Enactments of Expertise," *Annual Review of Anthropology* 39 (2010). The role of pathologization in medicine and science owes much of

its analysis to Canguilhem, *The Normal and the Pathological*. For a discussion of the role of imagination in knowledge production, see Vincent Crapanzano, *Imaginative Horizons: An Essay in Literary-Philosophical Anthropology* (Chicago: University of Chicago Press, 2003).

18. Charles Rosenberg, *Cholera Years: The United States in 1832, 1849, and 1866* (Chicago: University of Chicago Press, 1987 [1968]).

19. Warwick Anderson, *Colonial Pathologies: American Tropical Medicine, Race, and Hygiene in the Philippines* (Durham, N.C.: Duke University Press, 2006); David Arnold, *Colonizing the Body: State Medicine and Epidemic Disease in Nineteenth-Century India* (Berkeley: University of California Press, 1993).

20. For a discussion of epidemics and the geopolitics of quarantine, see Bruce Braun, "Biopolitics and the Molecularization of Life," *Cultural Geographies* 14, no. 1 (2007).

21. www.selectcomfort.com.

22. For anthropological approaches to compliance, see *Anthropology and Medicine* 17, no. 2 (2010).

23. In using *formation,* I am drawing on Deleuze and Guattari's idea of the "assemblage," which they describe as composed of four axes, namely, "a machinic assemblage of bodies, actions and passions, an intermingling of bodies reacting to one another; on the other hand it is a collective assemblage of enunciation, of acts and statements, of incorporeal transformations attributed to bodies. . . . the assemblage also has both territorial sides, or reterritorialized sides, which stabilize it, and cutting edges of deterritorialization, which carry it away." Deleuze and Guattari, *A Thousand Plateaus,* 88. This is their means of describing the relation between forms of expression and forms of content, which they explain as being necessarily variable: "There are variables of content, or proportions in the interminglings or aggregations of bodies, and there are variables of expression, factors internal to enunciation" (ibid.).

24. Deleuze and Guattari argue that "expression is independent [of content] and that this is precisely what enables it to react upon contents" (ibid., 89).

25. Lefebvre, "The Everyday and Everydayness."

26. Lefebvre, *Rhythmanalysis.*

27. Davis, *Enforcing Normalcy;* Michel Foucault, "Governmentality," trans. Robert Hurley, in *Power,* ed. James D. Faubion, vol. 3 of *Essential Works of Michel Foucault, 1954–1984* (New York: New Press, 2000 [1978]); Ian Hacking, *The Taming of Chance* (New York: Cambridge University Press, 1990); Ian Hacking, "How Should We Do the History of Statistics?" in *The Foucault Effect: Studies in Governmentality,* ed. Graham Burchell, Colin Gordon, and Peter Miller (Chicago: University of Chicago Press, 1991).

28. Davis, *Enforcing Normalcy.*

29. Foucault, *Discipline and Punish;* Michel Foucault, "The Subject and Power," trans. Robert Hurley, in *Power,* ed. Faubion.

30. In their précis of biopolitics, Ferenc Fehér and Agnes Heller make a similar remark regarding the function of "health": "By and large the politics of health is successful insofar as it transplants a massive guilt feeling, the prerequisite for the victory of the course of 'discipline and punish,' into the psyche of the individual." Ferenc Fehér and Agnes Heller, *Biopolitics: The Politics of the Body, Race and Nature,* vol. 15 of Public Policy and Social Welfare series, ed. Bernd Marin (Brookfield, U.K.: Avebury, 1994), 68.

31. In making this argument, I draw upon the work of Douglas R. Holmes, *Integral Europe: Fast-Capitalism, Multiculturalism, Neofascism* (Princeton, N.J.: Princeton University Press, 2000), 3–4. In a much different context, Holmes argues, "Integralism . . . [has] four registers: as a framework of meaning, as a practice of everyday life, as an idiom of solidarity, and, above all, as a consciousness of belonging linked to a specific cultural milieu. I also recognized that within these integralist practices were intriguing, though usually quiescent, struggles that under certain conditions could assume a volatile political character. Those who conjured this type of political insurgency drew on *adherents'* fidelity to specific cultural traditions and sought to recast these traditions within a distinctive historical critique and an exclusionary political economy. What seemed to catalyze this transformation was a broadly experienced rupture in the sense of belonging on the part of members of various communities and collectivities." Holmes bases his understanding of integralism in the lives of rural Italians and later European nationalist movements. This may seem distant from my concerns with medicine. However, as I explain, contemporary medicine has increasingly become a subjectivity-producing site, one that modern American desire is intimately caught up in. As medicine becomes an increasingly central means through which individuals come to know themselves and others, the identification of individuals with their treatments, and their being cared for by others and the state, becomes progressively contested ground for inclusion and exclusion. Whereas Holmes is interested in integralism based in culture and identity, I am interested in it as a logic for formations, which secondarily found cultural logics and identifications.

32. Peter Conrad, *The Medicalization of Society: On the Transformation of Human Conditions into Treatable Disorders* (Baltimore: Johns Hopkins University Press, 2007).

33. On the subject of citizenship and belonging mediated through scientific and medical classifications, see Deborah Heath, Rayna Rapp and Karen-Sue Taussig, "Genetic Citizenship," in *A Companion to the Anthropology of Politics,* ed. David Nugent and Joan Vincent (Malden, Mass.: Blackwell Publishers, 2005);

Rayna Rapp, Deborah Heath, and Karen-Sue Taussig, "Genealogical Dis-Ease: Where Hereditary Abnormality, Biomedical Explanation, and Family Responsibility Meet," in *Relative Values: Reconfiguring Kinship Studies,* ed. Sarah Franklin and Susan McKinnon (Durham, N.C.: Duke University Press, 2001); and Karen-Sue Taussig, Rayna Rapp, and Deborah Heath, "Flexible Eugenics: Technologies of the Self in the Age of Genetics," in *Genetic Nature/Culture: Anthropology and Culture beyond the Two Culture Divide,* ed. Amy Goodman, Deborah Heath, and Susan Lindee (Berkeley: University of California, 2003). For another perspective on the way that individuals come to identify as disordered and in need of medical mediation, see also Joseph Dumit, "Illnesses You Have to Fight to Get: Facts as Forces in Emergent, Uncertain Illnesses," *Social Science and Medicine* 62, no. 3 (2006).

34. For a discussion of the need for contextual relations for the production of meaning in any one domain, see Deleuze and Guattari, *A Thousand Plateaus,* 399.

35. Bruno Latour, *We Have Never Been Modern,* trans. Catherine Porter (Cambridge, Mass.: Harvard University Press, 1993 [1991]).

36. Michael Taussig, "Reification and the Consciousness of the Patient," *Social Science and Medicine* 14, no. 1 (1980).

37. Félix Guattari, *The Three Ecologies,* trans. Ian Pindar and Paul Sutton (New Brunswick, N.J.: Athlone, 2000 [1989]), 38.

38. Stiegler, *Technics and Time.*

39. Wolf-Meyer, "The Myth of Natural Sleep, or Technology and the Moral Authority of Primordial Thought."

40. For an overview, see David Harvey, *A Brief History of Neoliberalism* (New York: Oxford University Press, 2005).

41. Murray Melbin, *Night as Frontier: Colonizing the World after Dark* (New York: Free Press, 1987).

42. For a fictional representation of this, see Diana Gillon and Meir Gillon, *The Unsleep* (New York: Ballantine Books, 1961).

43. Karl Marx, *Early Writings,* trans. Rodney Livingstone and Gregor Benton (New York: Penguin, 1992 [1975]).

44. See, for three very different discussions of capitalism's focus on the whole body of the worker and its potentials for labor, Sidney Mintz, *Sweetness and Power* (New York: Penguin, 1985); Anson Rabinbach, *The Human Motor: Energy, Fatigue, and the Origins of Modernity* (Berkeley: University of California, 1990); and Edward Palmer Thompson, *The Making of the English Working Class* (New York: Penguin, 1980 [1963]). For a review of new levels of control and profit, see Stefan Helmreich, "Species of Biocapital," *Science as Culture* 17, no. 4 (2008).

45. Frederick Winslow Taylor, *The Principles of Scientific Management* (Mineola, N.Y.: Dover Publications, 1998 [1911]).

46. David Harvey, *The Condition of Postmodernity: An Enquiry into the Origins of Cultural Change* (Malden, Mass.: Blackwell, 1990).

47. Martin, *Flexible Bodies;* Juliet B. Schor, *The Overworked American: The Unexpected Decline of Leisure* (New York: Basic Books, 1991).

1. The Rise of American Sleep Medicine

1. Henry M. Lyman, *Insomnia and Other Disorders of Sleep* (Chicago: W. T. Keener, 1885), 1.

2. William A. Hammond, *Sleep and Its Derangements* (Philadelphia: J. B. Lippincott, 1869), 42.

3. Robert Macnish, *The Philosophy of Sleep* (New York: D. Appleton, 1824), 2.

4. Donald Laird, *How to Sleep and Rest Better* (New York: Funk and Wagnalls Company, 1937), 27–28.

5. M. A. Carskadon, W. C. Dement, M. M. Mitler, T. Roth, P. R. Westbrook, and S. Keenan, "Guidelines for the Multiple Sleep Latency Test (MSLT): A Standard Measure of Sleepiness," *Sleep* 9 (1986).

6. I discuss the goings-on at clinical meetings in more detail in Matthew Wolf-Meyer, "Sleep, Signification, and the Abstract Body of Allopathic Medicine," *Body and Society* 14, no. 4 (2008).

7. Sleep-related eating disorder (SRED) is a variation of sleepwalking behavior that results in the sleepwalker locating food, sometimes preparing it (such as soup), and eating it while asleep. High-fat-content foods are often sought out, making peanut butter one of the most frequently reported foods to be consumed by individuals experiencing SRED. SRED is often treated like other sleepwalking behaviors, with powerful sleep-inducing drugs in an attempt to consolidate nighttime sleep or with antidepressants, which also often consolidate sleep. SRED incidents increased with the popular use of Ambien and other modern sleep-inducing drugs as a result of biological and chemical conflicts during sleep. Tom Valeo, "Do Sleeping Pills Cause Night-Eating?" *Neurology Today* 6, no. 8 (2006).

8. William C. Dement and Christopher Vaughan, *The Promise of Sleep: A Pioneer in Sleep Medicine Explores the Vital Connection between Health, Happiness, and a Good Night's Sleep* (New York: Delacorte Press, 1999), 17.

9. Nathaniel Kleitman, *Sleep and Wakefulness* (Chicago: University of Chicago Press, 1963 [1933]), 3–4.

10. Sigmund Freud, *The Interpretation of Dreams,* trans. Joyce Crick (New York: Oxford University Press, 1999 [1901]).

11. Chip Brown, "The Stubborn Scientist Who Unraveled a Mystery of the Night," *Smithsonian,* October 2003; J. Allan Hobson, *Dreaming: An Introduction to the Science of Sleep* (New York: Oxford University Press, 2002), 42–44; Andrea Rock, *The Mind at Night: The New Science of How and Why We Dream* (New York: Basic Books, 2004), 1–16.

12. Macnish, *The Philosophy of Sleep.*

13. Eugene Aserinsky and Nathaniel Kleitman, "Regularly Occurring Periods of Eye Motility, and Concomitant Phenomena, during Sleep," *Science* 118 (1953).

14. Dement and Vaughan, *The Promise of Sleep,* 32–35.

15. Dreaming—in one form or another, again faltering on the subjective/objective split inherent in most dream research—has been shown to occur during different sleep phases. For a discussion, see D. Foulkes, "Dream Reports from Different Stages of Sleep," *Journal of Abnormal and Social Psychology* 65 (1962); D. Foulkes, "Nonrapid Eye Movement Mentation," *Experimental Neurology,* Supplement 4 (1967); Tore A. Nielsen, "A Review of Mentation in REM and NREM Sleep: 'Covert' REM Sleep as a Possible Reconciliation of Two Opposing Models," in *Sleep and Dreaming: Scientific Advances and Reconsiderations,* ed. Edward F. Pace-Schott, Mark Solms, Mark Blagrove, and Stevan Harnad (New York: Cambridge University Press, 2003); and Mark Solms, "Dreaming and REM Sleep Are Controlled by Different Brain Mechanisms," in *Sleep and Dreaming,* ed. Pace-Schott, Solms, Blagrove, and Harnad (New York: Cambridge University Press, 2003). These studies often argue that REM sleep and dreaming should be understood as separate phenomena, and, as such, a nonphysiological understanding of dreaming is necessary to explain their existence and evolutionary and social functions.

16. William Offenkrantz and Allan Rechtschaffen, "Clinical Studies of Sequential Dreams," *Archives of General Psychiatry* 8, no. 5 (1963); Allan Rechtschaffen, "Dream Reports and Dream Experiences," *Experimental Neurology,* Supplement 4 (1967); Allan Rechtschaffen, G. Vogel, and G. Shaikun, "Interrelatedness of Mental Activity during Sleep," *Archives of General Psychiatry* 9 (1963).

17. Dement and Vaughan, *The Promise of Sleep,* 71.

18. Ibid., 175–76.

19. Ibid., 176.

20. Mark Mahowald and Carlos Schenck, "Insights from Studying Human Sleep Disorders," *Nature* 437 (2005).

21. Quoted in Carlos H. Schenck, *Paradox Lost: Midnight in the Battleground of Sleep and Dreams* (Minneapolis: Extreme Nights LLC, 2005), 68.

22. Pasquale Montagna, Pierluigi Gambetti, Pietro Cortelli, and Elio Lugaresi, "Familial and Sporadic Fatal Insomnia," *Lancet Neurology* 2, no. 4 (2003): 168.

2. The Protestant Origins of American Sleep

1. Sacvan Bercovitch, *The American Jeremiad* (Madison: University of Wisconsin Press, 1978).

2. Cotton Mather, *The Serviceable Man: A Discourse Made Unto the General Court of the Massachusetts Colony, New-England, at the Anniversary Election 28d. 3m. 1690* (Massachusetts Colony: Joseph Browning, 1690), 55.

3. Cotton Mather, *A Midnight Cry an Essay for Our Awakening out of That Sinful Sleep, to Which We Are at This Time Too Much Disposed; and for Our Discovering of What Peculiar Things There Are in This Time, That Are for Our Awakening* (Boston: Samuel Phillips, 1692); Cotton Mather, *Vigilius: Or, the Awakener, Making a Brief Essay, to Rebuke First the Natural Sleep Which Too Often Proves a Dead Fly, in the Devotions of Them That Indulge It* (Boston: J. Franklin, 1719).

4. Cotton Mather, *The Wonders of the Invisible World: Observations as Well Historical as Theological, Upon the Nature, the Number, and the Operations of the Devils* (Boston: Benjamin Harris, 1693), 8.

5. Benjamin Franklin, "An Economical Project," in *Writings*, edited by J. A. Leo Lemay (New York: Library of America, 1987 [1784]), 984–85.

6. Ibid., 987.

7. William Whitty Hall, *Sleep* (New York: W. W. Hall, 1861).

8. Regarding the individualization of beds, see Lawrence Wright, *Warm and Snug: The History of the Bed* (Gloucestershire, U.K.: Sutton Publishing, 2004 [1962]); and Anthony Burgess, *On Going to Bed* (New York: Abbeville Press, 1982); on the manufacture of homes with rooms dedicated to individual occupants, see chapter 8 in Witold Rybczynski, *City Life* (New York: Simon and Schuster, 1995), and chapters 3 and 5 in R. Buckminster Fuller, *Nine Chains to the Moon* (Garden City, N.Y.: Anchor Books, 1971 [1963]).

9. Hall, *Sleep*, 19.

10. Ibid., 20.

11. Ibid., 115–16.

12. Ibid., 184–85.

13. Ibid., 192–93.

14. Ibid., 119.

15. Hammond, *Sleep and Its Derangements*, 14.

16. Ibid., 50.

17. Ibid., 13. Hammond explains this further: "The exciting cause of natural and periodic sleep is undoubtably to be found in the fact that the brain at stated times requires repose, in order that the cerebral substance which has been decomposed by mental and nervous action may be replaced by new material." Ibid., 18.

18. Ibid., 16.

19. Ibid., 43.

20. Ibid., 91.

21. Ibid., 47.

22. Ibid., 85.

23. Andrew Wilson, "The Ape of Death," *Harper's New Monthly Magazine,* no. 98, December 1898–May 1899, 782.

24. Ibid., 775.

25. Ibid., 774.

26. Ibid., 779.

27. Laird, *How to Sleep and Rest Better.*

28. Ibid., 21.

29. Ibid., 23.

30. Ibid., 38.

31. Ibid., 33–34.

32. Ibid., 80–81. Compare with Hugo Gernsback, "Learn and Work While You Sleep," *Science and Invention,* December 1921.

33. Laird, *How to Sleep and Rest Better,* 82.

34. I discuss this experiment in more detail in Matthew Wolf-Meyer, "Where Have All Our Naps Gone? or Nathaniel Kleitman, the Consolidation of Sleep, and Historiography of Emergence" (n.d.).

35. Nathaniel Kleitman Papers, series II, box 12, folder 12, Special Collections Research Center, University of Chicago Library.

36. This process of making the part, the individual, stand for the whole is the work of metonymy. However, in the construction of this relationship, there is an inevitable mismatch between the experiences of individuals and their expectations, as in the case of normative ideals generally; that is, individuals will never find that their experiences align with dominant expectations. For a discussion of this process in other contexts, see Homi Bhabha, *The Location of Culture* (New York: Routledge, 1994); and Michael Taussig, *Mimesis and Alterity: A Particular History of the Senses* (New York: Routledge, 1993).

37. Nathaniel Kleitman Papers, series II, box 10, folder 8.

38. Ibid.

39. Ibid.

40. Ibid., series II, box 11, folder 4.

41. Ibid., series II, box 10, folder 8.

42. Dement and Vaughan, *The Promise of Sleep,* 81.

43. Ibid., 98.

44. Ibid., 99.

45. Wolf-Meyer, "The Myth of Natural Sleep, or Technology and the Moral Authority of Primordial Thought."

46. Dement and Vaughan, *The Promise of Sleep*, 95.

47. Ibid., 71.

48. Ibid., 78.

49. Ibid., 92.

50. Ibid., 102.

51. Ibid., 15.

3. Sleeping and Not Sleeping in the Clinic

1. NSF Alert, November 10, 2004.

2. For a discussion of the rise of statistical and categorical thinking, see Hacking, *The Taming of Chance*.

3. This is an argument elaborated in Davis, *Enforcing Normalcy*.

4. See, for example, Gay Becker, "Deadly Inequality in the Health Care 'Safety Net': Uninsured Ethnic Minorities' Struggle to Live with Life-Threatening Illnesses," *Medical Anthropology Quarterly* 18, no. 2 (2004); and Roberta Bivins, *Alternative Medicine? A History* (New York: Oxford University Press, 2007).

5. Three illustrative sources include Karen Flint, *Healing Traditions: African Medicine, Cultural Exchange, and Competition in South Africa, 1820–1948* (Columbus: Ohio University Press, 2008); Jean M. Langford, *Fluent Bodies: Ayurvedic Remedies for Postcolonial Imbalance* (Durham, N.C.: Duke University Press, 2002); and Mei Zhan, *Other-Worldly: Making Chinese Medicine through Transnational Frames* (Durham, N.C.: Duke University Press, 2009).

6. For discussions of reification in medicine, see Martin, *The Woman in the Body;* Martin, *Flexible Bodies;* and Taussig, "Reification and the Consciousness of the Patient."

7. For discussions of how race becomes reified as the basis of disorder, see Jonathan Kahn, "How a Drug Becomes 'Ethnic': Law, Commerce, and the Production of Racial Categories in Medicine," *Yale Journal of Health Policy, Law, and Ethics* 4, no. 1 (2004); and Keith Wailoo, *Dying in the City of the Blues: Sickle Cell Anemia and the Politics of Race and Health* (Chapel Hill: University of North Carolina Press, 2001).

8. For a discussion of how biology lays the basis for modern understandings of individuals and the causation of their disorders, see Georges Canguilhem, *Ideology and Rationality in the History of the Life Sciences,* trans. Arthur Goldhammer (Cambridge, Mass.: MIT Press, 1990 [1977]); and Richard Lewontin, *Biology as Ideology: The Doctrine of DNA* (New York: Perennial, 1993).

4. Desiring a Good Night's Sleep

1. I use *intimacy* as synonymous with *capacity:* these alliances are about producing new capacities of bodies, new affects, through their interactions. For a discussion of affect and capacity, see Gilles Deleuze, *Spinoza: Practical Philosophy,* trans. Robert Hurley (San Francisco: City Lights Books, 1988 [1981]).

2. The literature on pharmaceuticals and personhood is extensive. For two surveys, see Emily Martin, "The Pharmaceutical Person," *BioSocieties* 1, no. 3 (2006); and Susan Reynolds Whyte, Sjaak Van der Geest, and Anita Hardon, *Social Lives of Medicines* (New York: Cambridge University Press, 2002).

3. Foucault, *Discipline and Punish;* Edward Palmer Thompson, *Customs in Common: Studies in Traditional Popular Culture* (New York: New Press, 1993).

4. Louis Althusser, *Lenin and Philosophy and Other Essays,* trans. Ben Brewster (New York: Monthly Review Press, 1971).

5. Bennett Weinberg and Bonnie Bealer, *The World of Caffeine: The Science and Culture of the World's Most Popular Drug* (New York: Routledge, 2001).

6. I refer to this crisis of desire and intimacy elsewhere as "pharmakological," see Matthew Wolf-Meyer, "Precipitating Pharmakologies and Capital Entrapments: Narcolepsy and the Strange Cases of Provigil and Xyrem," *Medical Anthropology: Cross-Cultural Studies in Health and Illness* 28, no. 1 (2009). My interest in the term *pharmakological* is that it highlights the doubtful context that treatment often produces, in addition to creating a new context that in turn entails its own treatment—sometimes social, sometimes chemical.

7. Mel Abel died in April 2004, at the very beginning of my fieldwork, so I was never able to interview him. My choice to draw on his account is due to how well articulated his experience of the disorder is, as well as his centrality in the generation of the nosological category: he was featured in the early articles on RBD, and at his death Mahowald remarked that it was "the end of an era," indicating Abel's importance to RBD research. Abel was also a popular case and was presented on CNN, ABC's *20/20,* and in the *New York Times Magazine,* February 2, 2003, in Chip Brown, "The Man Who Mistook His Wife for a Deer."

8. Quoted in Schenck, *Paradox Lost,* 66.

9. Ibid., 32.

10. Quoted ibid., 97, emphasis removed.

11. Quoted ibid., 97.

5. Now I Lay Me Down to Sleep

1. For an extensive bibliography of children's books, see Hillel Schwartz, *Making Noise: From Babel to the Big Bang and Beyond* (New York: Zone Books, 2011).

2. Margaret Wise Brown, *Goodnight Moon,* illustrated by Clement Hurd (New York: Harper and Brothers, 1990 [1947]).

3. www.cpsc.gov/. See also the original study by Suad Nakamura, Marilyn Wind, and Mary Ann Danello, "Review of Hazards Associated with Children Placed in Adult Beds," *Archives of Pediatric and Adolescent Medicine* 153 (October 1999).

4. www.washingtonpost.com/.

5. Elizabeth Pantley, *The No-Cry Sleep Solution: Gentle Ways to Help Your Baby Sleep through the Night* (New York: Contemporary Books, 2002), 38–39.

6. A more nuanced view is offered by Penelope Leach, *Your Baby and Child: From Birth to Age Five* (New York: Alfred A. Knopf, 1997 [1977]). Leach reviews the data for and against the "family bed" and argues that the "difference is a difference of evidence" (183). Her leanings are toward cosleeping, but she acknowledges the veracity of contrary opinions.

7. See, for example, James McKenna, *Researching the Sudden Infant Death Syndrome (SIDS): The Role of Ideology in Biomedical Sciences* (Stony Brook, N.Y.: State University of New York Press, 1991); and James McKenna, "Cultural Influences on Infant and Childhood Sleep Biology, and the Science That Studies It: Toward a More Inclusive Paradigm," in *Sleep and Breathing in Children: A Developmental Approach,* ed. Gerald Loughlin, John Carroll, and Carole Marcus (New York: Marcell Dakker, 2000).

8. Kleitman, *Sleep and Wakefulness,* 114–15.

9. Rosemary Wells, *Goodnight Max* (New York: Viking Children's, 2000). Like Rudy the Rooster, the protagonists of many of these children's books are animals, sometimes gendered through their names, but often genderless, allowing for the broadest possible spectrum of children to identify with them.

10. Mem Fox and Jane Dyer, *Time for Bed* (San Diego: Red Wagon Books, 1993).

11. Ibid., unpaginated, ellipsis in original.

12. Sylvia Long, *Hush Little Baby* (San Francisco: Chronicle Books, 1997).

13. Ibid., unpaginated. The original version reads as follows:

> Hush, little baby, don't say a word,
> Mama's gonna buy you a mockingbird.
> And if that mockingbird won't sing,
> Mama's gonna buy you a diamond ring.
> And if that diamond ring is brass,
> Mama's gonna buy you a lookingglass.
> And if that lookingglass gets broke,
> Mama's gonna buy you a nanny goat.
> And if that goat won't give no milk . . .

14. The anthropological literature on activities associated with children's sleep is actually quite diverse and detailed; however, there is little anthropological, or more properly "scientific," literature on actual sleep times from a cross-cultural perspective. Rather, most anthropological studies of children's sleep might be more properly understood as studies of ideal sleep times, placements, and positions. For a survey, see Judy DeLoache and Alma Gottlieb, *A World of Babies: Imagined Childcare Guides for Seven Societies* (New York: Cambridge University Press, 2000).

15. Martin Moore-Ede, *The Twenty-Four-Hour Society: Understanding Human Limits in a World That Never Stops* (Reading, Mass.: Addison-Wesley Publishing, 1993), 204.

16. Ibid., 6.

17. For discussions of temporal consolidation as it relates to nations and states, see Carol Greenhouse, *A Moment's Notice: Time Politics across Cultures* (Ithaca, N.Y.: Cornell University Press, 1996); David S. Landes, *Revolution in Time: Clocks and the Making of the Modern World* (Cambridge, Mass.: Belknap Press, 1983); and Wolfgang Schivelbusch, *The Railway Journey: The Industrialization of Time and Space in the 19th Century* (Berkeley: University of California Press, 1986 [1977]).

6. Pharmaceuticals and the Making of Modern Bodies and Rhythms

1. www.provigil.com/.

2. Provigil (modafinil) drug fact sheet, Cephalon, West Chester, Pa., 2004.

3. Michael J. Thorpy, "Sodium Oxybate for the Treatment of Narcolepsy," *Expert Opinion on Pharmacotherapy* 6, no. 2 (2005): 333.

4. Sebastian Overeem, Emmanuel Mignot, J. Gert van Dijk and Gert Jan Lammers, "Narcolepsy: Clinical Features, New Pathophysiologic Insights, and Future Perspectives," *Journal of Clinical Neurophysiology* 18, no. 2 (2001): 86.

5. Cited from *Sleep Hygiene*, a pamphlet produced by the American Academy of Sleep Medicine for distribution in sleep clinics.

7. Early to Rise

1. NSF, *2002 Sleep in America Poll* (Washington, D.C.: National Sleep Foundation, 2002), 7.

2. Ibid.

3. It has been widely claimed that because of the gendered roles of family members, with much of the caretaking of the family falling on women, women also deal with increased incidences of sleep disorders, particularly insomnia.

The relationship between insomnia and women's sleep was a primary focus of the National Sleep Foundation's 2007 Sleep in America poll, which took as its primary concern the changes in women's sleep throughout the life course. See NSF, *2007 Sleep in America Poll: Summary of Findings* (Washington, D.C.: National Sleep Foundation, 2007).

4. Jacques Le Goff, *Time, Work and Culture in the Middle Ages,* trans. Arthur Goldhammer (Chicago: University of Chicago Press, 1982).

5. See Christopher L. Drake, Timothy Roehrs, Gary Richardson, James K. Walsh, and Thomas Roth, "Shift Work Sleep Disorder: Prevalence and Consequences beyond That of Symptomatic Day Workers," *Sleep* 27, no. 8 (2004): 1459.

6. See, for example, John M. de Castro, "Circadian Rhythms of the Spontaneous Meal Pattern, Macronutrient Intake, and Mood of Humans," *Physiology and Behavior* 40, no. 4 (1987).

8. Chemical Consciousness

1. At the turn of the twenty-first century, the link between insomnia and depression had begun to be extensively researched, because insomnia was increasingly accepted as a symptom of depression and was often treated clinically by treating the depression rather than the insomnia as a primary concern. See Michel Billiard and Alison Bentley, "Is Insomnia Best Categorized as a Symptom or a Disease?" *Sleep Medicine* 5, Supplement 1 (2004).

2. Carlos H. Schenck, Scott R. Bundlie, Milton G. Ettinger, and Mark W. Mahowald, "Chronic Behavioral Disorders of Human REM Sleep: A New Category of Parasomnia," *Sleep* 9, no. 2 (1986).

9. Sleeping on the Job

1. See, for example, Harvey, *The Condition of Postmodernity.*

2. Quoted in Joanne Deschenaux, "Less Time for Lunch: The Siesta in Spain Is Disappearing under the Pressures of International Business and Big-City Commuting," *HR Magazine* 53, no. 6 (2008): 126.

3. The Fundación Independiente Web site, www.fundacionindependiente .es/, hosts a number of ideologically harmonious studies.

4. See Megan Brown, "Taking Care of Business: Self-Help and Sleep Medicine in American Corporate Culture," *Journal of Medical Humanities* 25, no. 3 (2004).

5. Pietro Basso, *Modern Times, Ancient Hours: Working Lives in the Twenty-First Century,* trans. Giacomo Donis (New York: Verso, 2003 [1998]), 75–76.

6. For a discussion of negotiating flexible work times, see Lonnie Golden, "Flexible Work Schedules: What Are We Trading Off to Get Them?" *Monthly Labor Review* 124, no. 3 (2001).

7. Mark Rosekind, Roy Smith, Donna Miller, Elizabeth Co, Kevin Gregory, Lissa Webbon, Philippa Gander, and J. Victor Lebacoz, "Alertness Management: Strategic Naps in Operational Setting," *Journal of Sleep Research* 4, Supplement 2 (1995).

8. William A. Anthony and Camille W. Anthony, "The Napping Company: Bringing Science to the Workplace," *Industrial Health* 43 (2005): 210.

9. Ibid., 211–12.

10. Aldous Huxley, *Brave New World* (New York: HarperCollins, 1998 [1932]).

11. The following data were provided by a FedEx/Kinko's employee in metro Detroit, who had worked for the company in a variety of managerial capacities since 1995 at the time of his interview in 2006.

10. Take Back Your Time

1. Juliet B. Schor, "The (Even More) Overworked American," in *Take Back Your Time: Fighting Overwork and Time Poverty in America,* ed. John de Graaf (San Francisco: Berrett-Koehler, 2003), 10–11.

2. www.timeday.org/reclaim_family_dinnertime.asp.

3. www.reclaimdinnertime.com/.

4. Unpublished handout.

5. Unpublished handout.

6. This summary is drawn from CAREI, *School Start Time Study: Final Report Summary* (Minneapolis: Center for Applied Research and Educational Improvement, 1998); and CAREI, *School Start Time Study: Technical Report,* vol. 2: *Analysis of Student Survey Data* (Minneapolis: Center for Applied Research and Educational Improvement, 1998).

7. Changes in adolescent sleep are summarized in Mary A. Carskadon, "When Worlds Collide: Adolescent Need for Sleep Versus Societal Demands," *Phi Delta Kappan* 80, no. 5 (1999).

8. For discussions of the development of American schooling, see R. Freeman Butts, *Public Education in the United States: From Revolution to Reform* (Austin: Holt, Rinehart and Winston, 1978); and Marvin Lazerson, *Origins of the Urban School: Public Education in Massachusetts, 1870–1915* (Cambridge, Mass.: Harvard University Press, 1971).

11. Unconscious Criminality

1. Charles Brockden Brown, *Edgar Huntly or, Memoirs of a Sleep-Walker* (New York: Penguin Books, 1988 [1799]), 83–85.

2. A *CSI: New York* episode portrayed a woman who dream enacted and was found in a suspicious pose (although asleep) over a corpse. This was enough for police to consider her a primary suspect in the corpse's murder, but she was eventually acquitted of any wrongdoing when it was discovered that her dream enactment was actually her nightly reliving of a traumatic experience from her past, and her posture over the corpse was wholly accidental. This was also the model adopted by Colin Wilson in *The Mammoth Book of True Crime* (New York: Carroll and Graf, 1998). Wilson discusses a handful of sleepwalking murder cases and attributes the behavior to either suppressed feelings or primordial survival instincts.

3. Brown, *Edgar Huntly or, Memoirs of a Sleep-Walker*, 152.

4. Edward van Every, *Sins of America (as "Exposed" by the Police Gazette)* (New York: Frederick A. Stokes Company, 1931).

5. Quoted in Samuel Gilman Brown, *The Works of Rufus Choate, with a Memoir of His Life* (Boston: Little, Brown and Company, 1862), 112.

6. http://hul.harvard.edu/huarc/refshelf/Sibley.htm#1846.

7. Discussed in D. Yellowlees, "Homicide by a Somnambulist," *Journal of Mental Science* 24 (1878).

8. For a full discussion of this case, see Shirley Frondorf, *Death of a "Jewish American Princess": The True Story of a Victim on Trial* (New York: Villard Books, 1988).

9. Quoted ibid., 145.

10. Mark Mahowald, S. R. Bundlie, T. D. Hurwitz, and Carlos Schenck, "Sleep Violence—Forensic Science Implications: Polygraphic and Video Documentation," *Journal of Forensic Sciences* 35, no. 2 (1990): 426.

11. See both R. Broughton, R. Billings, R. Cartwright, D. Doucette, J. Edmeads, M. Edward, F. Ervin, B. Orchard, R. Hill, and G. Turrell, "Homicidal Somnambulism: A Case Report," *Sleep* 17, no. 3 (1994); and June Callwood, *The Sleepwalker: The Trial That Made Canadian Legal History* (Toronto: Lester and Orpen Dennys, 1990).

12. Broughton et al., "Homicidal Somnambulism," 262.

13. Quoted ibid.

14. See the work conducted by Maurice Ohayon, Robert G. Priest, Jurgen Zulley, and Salvatore Smirne, "The Place of Confusional Arousals in Sleep and Mental Disorders: Findings in a General Population Sample of 13,057 Subjects,"

Journal of Nervous and Mental Disease 188, no. 6 (2000). Ohayon et al. argue that no violence is related to confusional arousals, although a number of other issues are related to confusional awakening.

15. Mahowald, Bundlie, Hurwitz, and Schenck, "Sleep Violence—Forensic Science Implications," 426.

16. Joshua Levine, "A Road to Injustice Paved with Good Intentions: Maggie's Misguided Crackdown on Drowsy Driving," *Hastings Law Journal* 56 (2005).

17. One of the first studies was conducted by Helinä Häkkänen and Heikki Summala, "Sleepiness at Work among Commercial Truck Drivers," *Sleep* 23, no. 1 (2000). This study evidences a high prevalence of microsleeps among truck drivers, who, as a population, experience chronic sleep debt.

18. The Wisconsin State Laboratory of Hygiene has conducted a number of studies on pharmaceuticals and their impairments to drivers. See Amy Cochems, Patrick Harding, and Laura Liddicoat, "Dextromethorphan in Wisconsin Drivers," *Journal of Analytical Toxicology* 31 (2007); and Laura Liddicoat and Patrick Harding, "Ambien: Drives Like a Dream?—Case Studies of Zolpidem-Impaired Drivers in Wisconsin," paper presented at the meeting of the American Academy of Forensic Sciences, Seattle, Wash., 2006.

19. Lawrence Block, *The Thief Who Couldn't Sleep* (New York: Fawcett Publications, 1984 [1966]), 9.

20. Ibid., 112.

21. Lawrence Block, *Me Tanner, You Jane* (New York: Fawcett Publications, 1986 [1970]), 9.

22. Lawrence Block, *Tanner's Twelve Swingers* (New York: Penguin, 1999 [1967]), 244–45.

23. Frederick Oughton, *The Two Lives of Robert Ledru: An Interpretive Biography of a Man Possessed* (London: Frederick Muller, 1963), 201–3; and Wilson, *The Mammoth Book of True Crime.*

12. The Extremes of Sleep

1. Joseph Bielitzki, *Enhancing Human Performance in Combat* (Arlington, Va.: Defense Sciences Office, 2002).

2. Ibid.

3. DARPA, *Fact File: A Compendium of DARPA Programs* (Arlington, Va.: Defense Advanced Research Projects Agency, 2003), 46–47.

4. Amy Kruse, *Defense and Biology: Fundamentals for the Future* (Arlington, Va.: Defense Sciences Office, 2005), 44–45.

5. Michel Siffre, *Beyond Time,* trans. Herma Briffault (New York: McGraw-Hill, 1964).

6. Ibid., 30–31.

7. Ibid., 57–58.

8. Ibid.

9. Ibid.

10. Ibid., 61.

11. Ibid., 103.

12. Ibid., 81.

13. Michel Siffre, "Six Months Alone in a Cave," *National Geographic,* March 1975, 228.

14. Siffre, *Beyond Time,* 217.

15. Ibid., 192.

16. Ibid., 208.

17. Recounted in Siffre, "Six Months Alone in a Cave."

18. www.poly-phasers.com/readarticle.php?article_id=17 (accessed August 2009).

19. Claudio Stampi and B. Davis, "Forty-Eight Days on the 'Leonardo Da Vinci' Strategy for Sleep Reduction: Performance Behaviour with Three Hours Polyphasic Sleep Per Day," *Sleep Research* 20 (1991).

20. http://news.bbc.co.uk/2/hi/uk_news/1180274.stm (last accessed May 22, 2007). See also Ellen MacArthur, *Taking on the World: A Sailor's Extraordinary Solo Race around the Globe* (Camden, Maine: International Marine/McGraw-Hill, 2003).

21. www.vendeeglobe.org/uk/magazine/2609.html (accessed March 3, 2007).

22. "When conditions allow, they sleep cycle by cycle, dive after dive, in other words in sequences lasting from 1 h 30 to 2 h on average, 3 times a day, twice at night, and once in early afternoon, at which time the body is most happy to sleep. In these conditions, the average length of a sleep is 5–6 hours, which is a minimum, but sufficient to avoid a lack of sleep. It is interesting to note that this ability to adapt to periods of sleep like this can be achieved in a few days. Back on dry land, it takes them sometimes several weeks to get back to a normal night of sleep. This short sleep technique cannot be used for more than a few days. Beyond that, you run the risk of falling asleep at any time and hallucinating. These very dangerous phenomena correspond to the falling asleep phase, where you find a mixture of real and subconscious thoughts. Falling asleep is only the start. It just requires an external stimulus to step back into reality and realise it was just a vision. Thus, sailors have imagined themselves back home or have seen a member of their family on board or talked to someone famous. Others have confused the sea with fields and waves with cows." www.vendeeglobe.org/uk/magazine/2935.html (accessed March 3, 2007).

23. Pete Goss, *Close to the Wind* (New York: Carroll and Graf, 1998), 167.

24. Ibid., 203.

25. Montagna, Gambetti, Cortelli, and Lugaresi, "Familial and Sporadic Fatal Insomnia," 167.

26. Jonathan D. Moreno, "Juicing the Brain," *Scientific American MIND* 17, no. 5, October 2006.

27. Quoted ibid.

Conclusion

1. David Pescovitz, ed., "The Future of Sleep," *Wired* 4, no. 12, December 1996.

2. Kleitman, *Sleep and Wakefulness,* 369.

3. Peretz Lavie, *The Enchanted World of Sleep,* trans. Anthony Berris (New Haven, Conn.: Yale University Press, 1998), 121.

4. For a discussion of new forms of social Darwinism, see Martin, *Flexible Bodies.*

5. Michel Jouvet, *The Paradox of Sleep: The Story of Dreaming,* trans. Laurence Garey (Cambridge, Mass.: MIT Press, 2001), 132–33.

6. Dement and Vaughan, *The Promise of Sleep,* 251–53.

7. Nancy Kress, *Beggars in Spain* (New York: Avon Books, 1993); Nancy Kress, *Beggars and Choosers* (New York: Tom Doherty Associates, 1994); Nancy Kress, *Beggars Ride* (New York: Tom Doherty Associates, 1996).

8. Gernsback, "Learn and Work While You Sleep."

9. For a discussion of "local" biologies, see Margaret Lock, *Encounters with Aging: Mythologies of Menopause in Japan and North America* (Berkeley: University of California Press, 1993).

Index

·················

MATTHEW J. WOLF-MEYER is assistant professor of anthropology at the University of California, Santa Cruz.